Navigation Control and Signal Processing Methods for Multiple Autonomous Unmanned Systems

Navigation Control and Signal Processing Methods for Multiple Autonomous Unmanned Systems

Guest Editors

Haoqian Huang
Bing Wang
Yuan Yang

MDPI

Basel • Beijing • Wuhan • Barcelona • Belgrade • Novi Sad • Cluj • Manchester

Guest Editors

Haoqian Huang
College of Artificial Intelligence and Automation
Hohai University
Changzhou
China

Bing Wang
College of Artificial Intelligence and Automation
Hohai University
Changzhou
China

Yuan Yang
School of Instrument Science and Engineering
Southeast University
Nanjing
China

Editorial Office
MDPI AG
Grosspeteranlage 5
4052 Basel, Switzerland

This is a reprint of the Special Issue, published open access by the journal *Applied Sciences* (ISSN 2076-3417), freely accessible at: https://www.mdpi.com/journal/applsci/special_issues/Multiple_Autonomous_Unmanned_Systems.

For citation purposes, cite each article independently as indicated on the article page online and as indicated below:

Lastname, A.A.; Lastname, B.B. Article Title. *Journal Name* **Year**, *Volume Number*, Page Range.

ISBN 978-3-7258-4803-4 (Hbk)
ISBN 978-3-7258-4804-1 (PDF)
https://doi.org/10.3390/books978-3-7258-4804-1

Contents

About the Editors . . . . **vii**

Preface . . . . **ix**

Haoqian Huang, Bing Wang and Yuan Yang
Navigation Control and Signal Processing Methods for Multiple Autonomous Unmanned Systems
Reprinted from: *Appl. Sci.* **2025**, *15*, 7335, https://doi.org/10.3390/app15137335 . . . . **1**

Guoqing Wu, Xiaoping Chen, Xuanyu Xie, Pu Zhang, Shenyu Ge, Wei Chen, et al.
Research on Hard, Transparent and Hydrophobic Coating on PMMA Sheet
Reprinted from: *Appl. Sci.* **2022**, *12*, 12927, https://doi.org/10.3390/app122412927 . . . . **9**

Mujie Zhao, Tao Zhang and Di Wang
A Novel UWB Positioning Method Based on a Maximum-Correntropy Unscented Kalman Filter
Reprinted from: *Appl. Sci.* **2022**, *12*, 12735, https://doi.org/10.3390/app122412735 . . . . **19**

Yu Sun, Bingbo Cui, Feng Ji, Xinhua Wei and Yongyun Zhu
The Full-Field Path Tracking of Agricultural Machinery Based on PSO-Enhanced Fuzzy Stanley Model
Reprinted from: *Appl. Sci.* **2022**, *12*, 7683, https://doi.org/10.3390/app12157683 . . . . **34**

Yinsheng Li, Bing Wang and Yuquan Chen
A Novel Decoupled Synchronous Control Method for Multiple Autonomous Unmanned Linear Systems: Bounded L_2-Gain for Coupling Attenuation
Reprinted from: *Appl. Sci.* **2022**, *12*, 7551, https://doi.org/10.3390/app12157551 . . . . **53**

Hanyu Sun, Guoqing Wu, Xueliang Wang, Tao Zhang, Pu Zhang, Wei Chen and Quanhua Zhu
Research on a Measurement Method for the Ocean Wave Field Based on Stereo Vision
Reprinted from: *Appl. Sci.* **2022**, *12*, 7447, https://doi.org/10.3390/app12157447 . . . . **66**

Zhuoran Xu, Manyi Wang, Qianmu Li and Linfang Qian
Fault Diagnosis Method Based on Time Series in Autonomous Unmanned System
Reprinted from: *Appl. Sci.* **2022**, *12*, 7366, https://doi.org/10.3390/app12157366 . . . . **86**

Haoqian Huang and Jiaying Wei
In-Motion Coarse Alignment Method Based on Position Loci and Optimal-REQUEST for SINS
Reprinted from: *Appl. Sci.* **2022**, *12*, 7113, https://doi.org/10.3390/app12147113 . . . . **103**

Yuxuan Zhao and Manyi Wang
The LOS/NLOS Classification Method Based on Deep Learning for the UWB Localization System in Coal Mines
Reprinted from: *Appl. Sci.* **2022**, *12*, 6484, https://doi.org/10.3390/app12136484 . . . . **121**

Jianqi Gao, Yanjie Li, Yunhong Xu and Shaohua Lv
A Two-Objective ILP Model of OP-MATSP for the Multi-Robot Task Assignment in an Intelligent Warehouse
Reprinted from: *Appl. Sci.* **2022**, *12*, 4843, https://doi.org/10.3390/app12104843 . . . . **139**

About the Editors

Haoqian Huang

Haoqian Huang received his bachelor's degree in automation from China University of Mining and Technology, Xuzhou, China, in 2007; his master's degree in agricultural mechanization engineering from Nanjing Agricultural University, Nanjing, China, in 2010; and his Ph.D. degree in instrument science and technology from Southeast University, Nanjing, in 2015. From 2015 to 2017, he was a Post-Doctoral Researcher at Southeast University. He then joined Hohai University, Nanjing, where he is an Associate Professor with the College of Artificial Intelligence and Automation. His main research interests include navigation technology applied to autonomous underwater vehicles, underwater sensing and positioning, and information fusion. As the first investigator, he has received multiple awards, including Third Prize at the Jiangsu Provincial Science and Technology Award, Second Prize at the China Instrument and Control Society Science and Technology Award, and Second Prize at the China Invention, Entrepreneurship and Innovation Award, and so on. He has served as the Director of China Inertial Technology Society, the Director of Jiangsu Instrument and Control Society, the Director of Jiangsu Oceanographic Society, Lead Guest Editor for a Special Issue of *Mathematical Problems in Engineering*, Guest Editor for a Special Issue of *Applied Sciences*, and others.

Bing Wang

Bing Wang received his Ph.D. in Control Theory and Control Engineering from the University of Science and Technology of China in 2006. He is a Professor and Doctoral Supervisor at the College of Artificial Intelligence and Automation, Hohai University. From 2009 to 2010, he was a Senior Visiting Scholar at the Department of Electrical Engineering and Computer Science, Case Western Reserve University, USA. Prof. Wang's research focuses on control theory for networked nonlinear systems and control technologies for renewable energy generation. He has led projects funded by the National Natural Science Foundation of China and has contributed to more than ten national and enterprise-level research initiatives. He serves as a Director of the Jiangsu Society of Automation, a Standing Member of the Youth Working Committee of the Chinese Association of Automation, and a Senior Member of China Electrotechnical Society.

Yuan Yang

Yuan Yang received her undergraduate degree in Communication Engineering in 2007 and her master's in Signal and Information Systems Engineering in 2010. In 2015, she earned her Ph.D. in Informatics from the School of Computer Science at the Free University of Berlin, Germany. Since August 2015, she has served as an Associate Professor and Master's student tutor at the School of Instrument Science and Engineering, Southeast University. Prof. Yang's research focuses on positioning signal modeling, data processing, and position control. Her work contributes to advancements in precision instrumentation and location-based technologies, combining communication systems and informatics to support innovations in modern engineering applications.

Preface

Autonomous unmanned systems are important in industry, for ocean-related activities, in agriculture, and so on. They are used at sea, on land, and in the air and play a significant role in all our lives. Multiple autonomous unmanned systems have become widely used and hot research topics due to their advantages. Navigation control and signal processing methods, as the key technologies for multiple autonomous systems, are the main focus of this Reprint. Some new methods of navigation and control, used in different autonomous systems, are introduced, while some novel signal processing methods also are reported. Dr. Haoqian Huang was mainly responsible for the original draft preparation, review, and editing for all manuscripts. Dr. Bing Wang and Dr. Yuan Yang mainly reviewed and edited manuscripts. This Reprint is intended as a reference for researchers, engineers, and other scholars in related fields.

Haoqian Huang, Bing Wang, and Yuan Yang
Guest Editors

applied sciences

Editorial

Navigation Control and Signal Processing Methods for Multiple Autonomous Unmanned Systems

Haoqian Huang [1,*], Bing Wang [1] and Yuan Yang [2]

[1] College of Artificial Intelligence and Automation, Hohai University, Changzhou 213200, China; icekingking@hhu.edu.cn
[2] School of Instrument Science and Engineering, Southeast University, Nanjing 210096, China; yangyuancsi@163.com
* Correspondence: hqhuang@hhu.edu.cn

1. Introduction

Autonomous underwater vehicles (AUVs) have been widely deployed in numerous underwater applications, such as biological monitoring, oceanographic surveys, and military underwater surveillance [1–4]. The localization of AUVs has usually been a challenge because of rapid attenuation of the radio frequency, such as global positioning system (GPS) signals. The high-precision accelerometer- and gyroscope-based inertial navigation system (INS) is the most common method of AUV navigation. Nevertheless, the navigation errors of AUVs accumulate over time because of the inertial sensors' drifts [5–8], which may cause unlimited increases in the localization error. The increase in localization errors can be corrected and updated by the GPS signal after surfacing, but it is impractical to surface when AUVs are working during a deep-water mission [9–11].

In the navigation of AUVs, a significant problem is determining the positions of AUVs by using a nonlinear filter. Nonlinear filtering is an effective method for inferring the unknown state of a nonlinear system and has been widely employed in a number of applications involving the navigation of AUVs, signal processing, communications, target tracking, and control [12–19]. A few nonlinear filters for underwater navigation have also been presented to improve the estimation accuracy, such as the extended Kalman filter (EKF), unscented Kalman filter (UKF), and cubature Kalman filter (CKF) [20–25].

The performance of the above filtering algorithms depends largely on a priori knowledge of the noise covariance matrix, and the use of an inaccurate noise covariance matrix can result in substantial estimation errors or even filtering divergence. However, in the navigation of AUVs, an accurate noise covariance matrix is very difficult to establish and may be time-varying, because the performance of sensors may vary with changes in the environment, which will degrade the navigation accuracy [26–29]. The adaptive algorithm is an effective method for solving the inherent problem of unknown noise covariance matrices in the underwater navigation of AUVs [30–35].

In contrast to fully actuated AUVs, underactuated AUVs are equipped with fewer propellers. This configuration offers distinct advantages and unique characteristics. The primary benefit of underactuated AUVs lies in their ability to perform intricate motions and maneuvers using fewer control inputs or actuators. As a result, underactuated AUVs offer simpler and more cost-effective design solutions [36–38]. However, due to their complicated interactions with the underwater environment, underactuated AUVs exhibit inherently nonlinear dynamics. Consequently, controlling such systems requires advanced control techniques to handle nonlinearities and ensure the stability of system.

From the above analysis, it can be seen that it is easy to be influenced by external disturbances to the nonlinearity characteristics, which brings obvious challenges to the controller design process [39–41]. To ensure the reliable operation of an AUV, it is crucial to model disturbances accurately and account for their effects. DO has gained widespread popularity in the field of control due to its superior performance in estimating uncertainties [42,43]. A DO was developed in [44] to assess the influence of disturbances while optimizing communication resource consumption, and intermittent measurement-based uncertainty was also considered, while the converge speed of estimation errors was ignored. In [45], a novel finite-time DO was proposed, but the upper bound of the convergence time is unknown. In [46], a fixed-time DO was proposed to determine the disturbance, and it was shown that the estimation errors can reach the zero domain within the settling time. An observer-based fixed-time sliding mode control (SMC) was proposed in [47], where a weak chattering DO was designed to estimate uncertainties within a fixed time. However, in most previous works, the upper bound of the disturbance is assumed to be known [48], which is not reasonable under specific conditions. By integrating the DO with the formation controller, this paper proposes a novel approach for underactuated AUV formation control.

Formation control strategies play a pivotal role in the field of unmanned systems, providing theoretical and practical frameworks for coordination and collaboration among multiple unmanned vehicles. Presently, most existing formation control strategies can be roughly categorized into leader–follower, behavior-based, virtual structure, graph theory-based, artificial potential function, and visually based strategies [49]. Among these, the "leader–follower" strategy stands out as one of the most often used and fundamental approaches in formation control. Under this strategy, one or more unmanned vehicles are designated as the leader, whose trajectory serves as a reference for other followers. The leader's movements determine the overall shape and behavior of the formation. Communication and mutual perception between the leader and followers are crucial in maintaining tight coordination within the formation.

The primary challenges associated with controlling underactuated AUV systems can be broadly categorized into three groups [50]. Firstly, in the existing literature [51,52], to simplify the process of controller design, the mass and damping matrices in underactuated AUV models are assumed to be diagonal, while this kind of model cannot reflect the actual conditions due to the inherent lack of symmetry between the bow and stern of the AUV. To overcome this limitation, ref. [53] introduced a coordinate transformation of the design controller. This conventional control design for such models often involves introducing a coordinate transformation which can transform ship dynamics to a "diagonal form" before addressing the coupling effects between the position and velocity variables, which can be cumbersome and time-consuming. Secondly, it is observed that the majority of existing formation control schemes, employing the leader–follower method, rely on velocity information from the leader [54,55]. However, in practical scenarios, measuring the velocity of the leader is challenging for followers due to noise contamination and communication delays. As a result, a more practical and constructive approach for formation control is to rely solely on measurements of positions and orientations, eliminating the need for velocity information [56,57]. Thirdly, an essential performance criterion for formation control is the converge speed. The fixed-time stability technique [58–60] has emerged as a valuable approach for designing a controller which can ensure a guaranteed converging time for the system, irrespective of the initial conditions of the agents. While the concept of fixed-time formation control holds promise, current research efforts have predominantly focused on addressing the consensus problem in general multi-agent systems and fully actuated AUVs, leaving room for further investigation and development in this specific area.

2. An Overview of the Published Articles

This section presents a comprehensive overview of the Special Issue title "Navigation Control and Signal Processing Methods for Multiple Autonomous Unmanned Systems", summarizing the key contributions of the published articles.

Multi-robot task assignment is one of the main processes in an intelligent warehouse. One of the papers models multi-robot task assignment in an intelligent warehouse as an open-path multi-depot asymmetric traveling salesman problem (OP-MATSP). A two-objective integer linear programming (ILP) model for solving OP-MDTSP is proposed. The theoretical bound on the computational time complexity of this model is O(n!). We can solve the small multi-robot task assignment problem by solving the two-objective ILP model using the Gurobi solver. The multi-chromosome coding-based genetic algorithm has a smaller search space, so we use it to solve large-scale problems. The experimental results reveal that the two-objective ILP model is very good at solving small-scale problems. For large-scale problems, both the EGA and NSGA3 genetic algorithms can efficiently obtain suboptimal solutions. This demonstrates that the paper's multi-robot work assignment methods are helpful in an intelligent warehouse [61].

Additionally, a localization system is one of the basic requirements for coal mines. Ultra-wideband (UWB), as a technology with broad application prospects, is considered to have great potential in the absence of satellite signals, especially in an underground mine environment, as it can provide rescue assistance. However, state-of-the-art UWB position systems in coal mines cannot efficiently differ the line of sight from all communication links, which results in the deterioration of the localization accuracy. In another paper in this Special Issue, the authors propose an LOS/NLOS classification method based on a deep learning algorithm. Specifically, they utilize the Generative Adversarial Network (GAN) to generate diagnostic data for frame transmission under non-line-of-sight (NLOS) conditions. Then, a Convolutional Neural Network (CNN) is adopted to identify NLOS communication. Finally, the trilateral centroid positioning algorithm (TCPA), based on ranging data, is used to verify the effect of our method for a localization system in coal mines [62].

Another recent study, included in this Special Issue, proposes an improved in-motion coarse alignment method for a strapdown inertial navigation system (SINS) using position loci obtained from the Global Positioning System (GPS). The difference from the popular coarse alignment methods is that the proposed algorithm uses GPS position loci information to form the vector observation and does not need velocity information, which expands the application range of in-motion coarse alignment. In addition, this paper utilizes the Optimal-REQUEST algorithm to reduce the influence of random errors contained in the vector observation. The Optimal-REQUEST algorithm is an adaptive iterative updating algorithm, which can adaptively adjust the gain of the filter according to the loss function. The simulation results confirmed that the proposed algorithm can suppress the impact of random errors effectively. The pitch, roll, and yaw angles calculated by the proposed algorithm were improved by 51.95%, 53.80%, and 63.03% compared with the comparison algorithms [63].

There are various types of autonomous unmanned systems, covering different areas of sea, land, and air and exhibiting significance in multiple fields of national security and social life. Due to the development of technology, the scale of unmanned systems is becoming increasingly large, the number of components in the system is increasing, and the operating environment of the system is becoming increasingly complex. Therefore, the probability of failure of the components of the system will also be significantly increased. In order to eliminate the impact of a fault in time, the fault diagnosis method is significant. Considering the differences in components in autonomous unmanned systems, if a specific fault diagnosis algorithm is designed for each type of component, it will lead to difficulties in the coordinated control of the system. Therefore, another paper analyzes the data characteristics of unmanned autonomous system devices (such as sensors) and finds that these data have time series. Therefore, the data for different devices can be converted into time series, and a general fault diagnosis algorithm that is suitable for most devices can be developed. The fault diagnosis algorithm is based on the clustering algorithm. In order to improve the clustering effect, the time series of different devices are represented by Gaussian mixture clustering to reduce the computational complexity of the clustering calculation. Then, a time series similarity measurement method based on the improved Markov chain is proposed. This method can distinguish normal samples from abnormal samples more successfully in order to classify and identify faults effectively [64].

The wave parameter is an important environmental input condition. Traditional contact wave measurement methods are unable to meet the requirements of high precision, non-contact, and enabling ship wave field assessment. Alternatively, stereo vision technology can realize a non-contact and mobile form of measurement. However, this technology suffers from poor efficiency and adaptability. Another paper in this Special Issue proposes a comprehensive wave measurement method that is based on stereo vision, wherein the gridding of siftGPU is used to achieve the fast matching of large images. The whole algorithm can be run within 6 s and guarantees more than 20,000 feature-matching logarithms. Furthermore, by utilizing the least squares method and sea surface wave surface theory, the sea surface base level can be calculated without control points, along with the inversion of the sea wave parameters (wave height, period, and wave direction) and error point fitting. The rationality and superiority of the algorithm were verified through multiple comparison experiments [65].

Another study addresses the distributed optimal decoupling synchronous control of multiple autonomous unmanned linear systems (MAUSs) that are subject to complex network dynamic coupling. The leader–follower mechanism based on neighborhood error dynamics is established, and the network coupling term is regarded as the external disturbance to realize the decoupling cooperative control of each agent. The Bounded L2-Gain problem for the network coupling term is formulated into a multi-player zero-sum differential game. It is shown that the solution to the multi-player zero-sum differential game requires the solution to the coupled Hamilton–Jacobi (HJ) equations. The coupled HJ equations are transformed into an algebraic Riccati equation (ARE), which can be solved to obtain the Nash equilibrium of a multi-player zero-sum game. It is shown that the bounded L2-Gain for coupling attenuation can be realized by applying the zero-sum game solution as the control protocol, and the ultimately uniform boundedness (UUB) of a local neighborhood error vector under conservative conditions is proved. A simulation example is provided to show the effectiveness of the proposed method [66].

The unmanned operation of agriculture machinery in a full farmland field is an important part of unmanned farm and smart agriculture. Although the autonomous navigation of agricultural robots has been widely studied in the literature, research on the full-field path tracking problem of agriculture machinery is rare. In a paper in this Special Issue, in order to enhance the adaptivity of the path tracking algorithm, an improved fuzzy Stanley model (SM) is proposed based on particle swarm optimization (PSO), where the control gain is modified adaptively according to the tracking error, velocity, and steering actuator saturation. The PSO-enhanced fuzzy SM (PSO-FSM) is verified by experiments on numerical simulation and the self-driving of a mobile vehicle. The simulation results indicate that the PSO-FSM achieves a better result than SM and FSM, where the PSO-FSM changes the control gain adaptively under different velocities and actuator saturation conditions, and the maximum lateral errors of the SM and PSO-FSM for mobile vehicle autonomous turning are 0.32 m and 0.03 m, respectively. When the location of the mobile vehicle deviates from the expected path at 4 m in a lateral direction, the distance of the guided trajectory for the mobile vehicle to reach the expected path is no more than 5 m [67].

In another study, to solve the problems of measurement information abnormal error and nonlinear filtering in UWB navigation and positioning, an ultra-wideband position algorithm based on a maximum cross-correlation entropy unscented Kalman filter is proposed. The algorithm first obtains the predictive state estimate and the covariance matrix through traceless transformation. Then, it reconstructs observation information using the nonlinear regression method based on the maximum cross-correlation entropy criterion, which enhances the robustness of the unscented Kalman filter algorithm for heavy-tailed noise. The simulation and actual test results show that this algorithm has better positioning accuracy and stability than the traditional filter algorithm in a non-Gaussian noise environment. This algorithm effectively solves the problem that the UWB indoor location is easily affected by indoor environments, resulting in fixed deviation for that location [68].

These studies collectively highlight the future of navigation control and signal processing methods for autonomous unmanned systems. Given the breadth of this topic, this Special Issue includes papers from focused research to broad overviews. It is clear that many factors, from cost-effectiveness and the role of people to the importance of different models, will influence the future of this field.

3. Conclusions

Navigation, guidance, control, and signal processing are important for a number of autonomous unmanned systems in ocean exploring, surveying, and so on. However, there are some challenges for autonomous unmanned systems due to the complex underwater environment. The articles published in this Special Issue mainly introduce and discuss how to solve the problems of navigation, control, and signal processing.

Future research should focus on the improvement of navigation, control, and positioning for multiple autonomous unmanned vehicles. Moreover, more state-of-the-art methods such as artificial intelligence, deep learning, and machine learning should be introduced and integrated with traditional navigation and control methods.

Author Contributions: H.H.: writing—original draft preparation; H.H., B.W. and Y.Y.: writing—review and editing. All authors have read and agreed to the published version of the manuscript.

Funding: This research received no external funding.

Conflicts of Interest: The authors declare no conflicts of interest.

References

1. Curtin, T.; Bellingham, J.; Catipovic, J.; Webb, D. Autonomous oceanographic sampling network. *Oceanography* **1993**, *6*, 86–94. [CrossRef]
2. Fallon, M.F.; Papadopoulos, G.; Leonard, J.J.; Patrikalakis, N.M. Cooperative AUV navigation using a single maneuvering surface craft. *Int. J. Robot. Res.* **2010**, *29*, 1461–1474. [CrossRef]
3. Jenkins, S.A.; D'Spain, G. Autonomous underwater glider. In *Springer Handbook of Ocean Engineering*; Springer: Dordrecht, The Netherlands; Heidelberg, Germany; London, UK; New York, NY, USA, 2016.
4. Tan, H.P.; Diamant, R.; Seah, W.K.G.; Waldmeyer, M. A survey of techniques and challenges in underwater localization. *Ocean Eng.* **2011**, *38*, 1663–1676. [CrossRef]
5. Wang, X.H.; Shirinzadeh, B.; Ang, M.H. Nonlinear double-integral observer and application to quadrotor aircraft. *IEEE Trans. Ind. Electron.* **2015**, *62*, 1189–1200. [CrossRef]
6. Rogne, R.H.; Bryne, T.H.; Fossen, T.I.; Johansen, T.A. Redundant MEMS-based inertial navigation using nonlinear observers. *J. Dyn. Syst. Meas. Control.* **2018**, *140*, 071001. [CrossRef]

7. Huang, Y.L.; Zhang, Y.G.; Xu, B.; Wu, Z.M.; Chambers, J.A. A new adaptive extended Kalman filter for cooperative localization. *IEEE Trans. Aerosp. Electron. Syst.* **2018**, *54*, 353–368. [CrossRef]
8. Huang, Y.L.; Zhang, Y.G.; Wang, X.D. Kalman-filtering-based inmotion coarse alignment for odometer-aided SINS. *IEEE Trans. Instrum. Meas.* **2017**, *66*, 3364–3377. [CrossRef]
9. Gao, W.; Liu, Y.L.; Xu, B. Robust Huber-based iterated divided difference filtering with application to cooperative localization of autonomous underwater vehicles. *Sensors* **2014**, *14*, 24523–24542. [CrossRef]
10. Yao, Y.Q.; Xu, X.S.; Yang, D.R.; Xu, X. An IMM-UKF aided SINS/USBL calibration solution for underwater vehicles. *EEE Trans. Veh. Technol.* **2020**, *69*, 3740–3747. [CrossRef]
11. Li, J.C.; Gao, W.; Zhang, Y. Gravitational apparent motion-based SINS self-alignment method for underwater vehicles. *IEEE Trans. Veh. Technol.* **2018**, *67*, 11402–11410. [CrossRef]
12. Huang, Y.L.; Zhang, Y.G.; Li, N.; Chambers, J. Robust student's t based nonlinear filter and smoother. *IEEE Trans. IEEE Trans. Aerosp. Electron. Syst.* **2016**, *52*, 2586–2596. [CrossRef]
13. Webster, S.E.; Eustice, R.M.; Singh, H.; Whitcomb, L.L. Advances in single-beacon one-way-travel-time acoustic navigation for underwater vehicles. *Int. J. Robot. Res.* **2012**, *31*, 935–950. [CrossRef]
14. Bahr, A.; Leonard, J.J.; Fallon, M.F. Cooperative localization for autonomous underwater vehicles. *Int. J. Robot. Res.* **2009**, *28*, 714–728. [CrossRef]
15. Zhang, Y.G.; Huang, Y.L.; Li, N.; Zhao, L. Interpolatory cubature Kalman filters. *IET Control Theory Appl.* **2015**, *9*, 1731–1739. [CrossRef]
16. Huang, Y.; Zhang, Y. Robust student's t-based stochastic cubature filter for nonlinear systems with heavy-tailed process and measurement noises. *IEEE Access* **2017**, *5*, 7964–7974. [CrossRef]
17. Ruiz, T.; De Raucourt, S.; Petillot, Y.; Lane, D.M. Concurrent mapping and localization using sidescan sonar. *IEEE J. IEEE J. Ocean. Eng.* **2004**, *29*, 442–456. [CrossRef]
18. Huang, Y.L.; Zhang, Y.G.; Wang, X.X.; Zhao, L. Gaussian filter for nonlinear systems with correlated noises at the same epoch. *Automatica* **2015**, *60*, 122–126. [CrossRef]
19. Zhang, Y.G.; Huang, Y.L.; Li, N.; Zhao, L. Embedded cubature Kalman filter with adaptive setting of free parameter. *Signal Process.* **2015**, *114*, 112–116. [CrossRef]
20. Julier, S.; Uhlmann, J.; Whyte, H.F.D. A new method for the nonlinear transformation of means and covariances in filters and estimators. *IEEE Trans. Autom. Control* **2000**, *45*, 477–482. [CrossRef]
21. Arasaratnam, I.; Haykin, S. Cubature Kalman filters. *IEEE Trans. Autom. Control* **2009**, *54*, 1254–1269. [CrossRef]
22. Huang, H.Q.; Zhou, J.; Zhang, J.; Yang, Y.; Song, R.; Chen, J.; Zhang, J. Attitude estimation fusing quasi-Newton and cubature Kalman filtering for inertial navigation system aided with magnetic sensors. *IEEE Access* **2018**, *6*, 28755–28767. [CrossRef]
23. Huang, H.Q.; Shi, D.R.; Zhou, J.; Yang, Y.; Song, R.; Chen, J.F.; Wu, G.Q.; Zhang, J.J. Attitude determination method integrating square-root cubature Kalman filter with expectation-maximization for inertial navigation system applied to underwater glider. *Rev. Sci. Instrum. Rev. Sci. Instrum.* **2019**, *90*, 095001. [CrossRef]
24. Huang, H.Q.; Chen, X.Y.; Zhou, Z.K.; Liu, H.; Lv, C.P. Study on INS/DR integration navigation system using EKF/RK4 algorithm for underwater gliders. *J. Mar. Sci. Technol.* **2017**, *25*, 84–95.
25. Huang, H.Q.; Chen, X.Y.; Zhang, B.; Wang, J. High accuracy navigation information estimation for inertial system using the multi-model EKF fusing adams explicit formula applied to underwater gliders. *ISA Trans.* **2017**, *66*, 414–424. [CrossRef] [PubMed]
26. Idkhajine, L.; Monmasson, E.; Maalouf, A. Fully FPGA-based sensorless control for synchronous AC drive using an extended Kalman filter. *IEEE Trans. Ind. Electron.* **2012**, *59*, 3908–3918. [CrossRef]
27. Yu, M.J. INS/GPS integration system using adaptive filter for estimating measurement noise variance. *IEEE Trans. Aerosp. Electron. Syst.* **2012**, *48*, 1786–1792. [CrossRef]
28. Fox, C.W.; Roberts, S.J. A tutorial on variational Bayesian inference. *Artif. Intell. Rev.* **2012**, *38*, 85–95. [CrossRef]
29. Li, X.R.; Bar-Shalom, Y. Multiple-model estimation with variable structure. *IEEE Trans. Autom. Control* **1996**, *41*, 478–493.
30. Dong, P.; Jing, Z.L.; Leung, H.; Shen, K. Variational Bayesian adaptive cubature information filter based on Wishart distribution. *IEEE Trans. Autom. Control* **2017**, *62*, 6051–6057. [CrossRef]
31. Kim, J.S.; Yum, D.H.; Lee, P.J.; Hong, S.J.; Kim, J. A hybrid sonar algorithm for submersible cars to verify the location of senders. *IEEE Trans. Veh. Technol.* **2012**, *61*, 2707–2714. [CrossRef]
32. Jagannathan, S.; Galan, G. One-layer neural-network controller with preprocessed inputs for autonomous underwater vehicles. *IEEE Trans. Veh. Technol.* **2003**, *52*, 1342–1355. [CrossRef]
33. Ma, Y.N.; Gong, Y.J.; Xiao, C.F.; Gao, Y.; Zhang, J. Path planning for autonomous underwater vehicles: An ant colony algorithm incorporating alarm pheromone. *IEEE Trans. Veh. Technol.* **2019**, *68*, 141–154. [CrossRef]
34. Cheng, X.Z.; Shu, H.N.; Liang, Q.H.; Du, D.H.C. Silent positioning in underwater acoustic sensor networks. *IEEE Trans. Veh. Technol.* **2008**, *57*, 1756–1766. [CrossRef]

35. Isbitiren, G.; Akan, O.B. Three-dimensional underwater target tracking with acoustic sensor networks. *IEEE Trans. Veh. Technol.* **2011**, *60*, 3897–3906. [CrossRef]
36. Huang, H.Q.; Tang, J.C.; Liu, C.; Zhang, B.; Wang, B. Variational Bayesian-Based Filter for Inaccurate Input in Underwater Navigation. *IEEE Trans. Veh. Technol.* **2021**, *70*, 8441–8452. [CrossRef]
37. Zhao, R.; Xu, J.; Xiang, X. A Review of Path Planning and Cooperative Control for MAUV Systems. *Chin. J. Ship Res.* **2018**, *13*, 58–65.
38. Khalil, H.K. *Nonlinear Control*; Pearson: New York, NY, USA, 2015.
39. Jiang, Z.P. Global Tracking Control of Underactuated Ships by Lyapunov's Direct Method. *Automatica* **2002**, *38*, 301–309. [CrossRef]
40. Aguiar, A.P.; Hespanha, J.P. Trajectory-Tracking and PathFollowing of Underactuated Autonomous Vehicles with Parametric Modeling Uncertainty. *IEEE Trans. Autom. Control* **2007**, *52*, 1362–1379. [CrossRef]
41. Cui, R.; Zhang, X.; Cui, D. Adaptive Sliding-Mode Attitude Control for Autonomous Underwater Vehicles with Input Nonlinearities. *Ocean Eng.* **2016**, *123*, 45–54. [CrossRef]
42. Liu, Y.C.; Liu, S.Y.; Wang, N. Fully-Tuned Fuzzy Neural Network based Robust Adaptive Tracking Control of Unmanned Underwater Vehicle with Thruster Dynamics. *Neurocomputing* **2016**, *196*, 1–13. [CrossRef]
43. Weng, Y.P.; Wang, N. SMC-based model-free tracking control of unknown autonomous surface vehicles. *ISA Trans.* **2022**, *130*, 684–691. [CrossRef] [PubMed]
44. Gao, Z.Y.; Guo, G. Fixed-Time Sliding Mode Formation Control of AUVs Based on a Disturbance Observer. *IEEE/CAA J. Autom. Sin.* **2020**, *7*, 539–545. [CrossRef]
45. Zhang, J.Q.; Yu, S.; Wu, D.F.; Yan, Y. Nonsingular Fixed-Time Terminal Sliding Mode Trajectory Tracking Control for Marine Surface Vessels with Anti-Disturbances. *Ocean Eng.* **2020**, *217*, 108158. [CrossRef]
46. Shi, Y.; Xie, W.; Zhang, G.; Dong, H.; Zhang, W. Event-Triggered Saturation-Tolerant Control for Autonomous Underwater Vehicles with Quantitative Transient Behaviors. *IEEE Trans. Veh. Technol.* **2023**, *72*, 9857–9867. [CrossRef]
47. Wang, N.; Qian, C.; Sun, J.C.; Liu, Y.C. Adaptive Robust FiniteTime Trajectory Tracking Control of Fully Actuated Marine Surface Vehicles. *IEEE Trans. Control Syst. Technol.* **2016**, *24*, 1454–1462. [CrossRef]
48. Van, M.; Ceglarek, D. Robust Fault Tolerant Control of Robot Manipulators with Global Fixed-Time Convergence. *J. Frankl. Inst.* **2021**, *358*, 699–722. [CrossRef]
49. Yang, Y.; Yang, X.; Li, T. A Survey of Autonomous Underwater Vehicle Formation: Performance, Formation Control, and Communication Capability. *IEEE Commun. Surv. Tutor.* **2021**, *23*, 815–841. [CrossRef]
50. Das, B.; Subudhi, B.; Pati, B.B. Cooperative Formation Control of Autonomous Underwater Vehicles: An Overview. *Int. J. Autom. Comput.* **2016**, *13*, 199–255. [CrossRef]
51. Antonelli, G.; Chiaverini, S.; Sarkar, N.; West, M. Adaptive Control of an Autonomous Underwater Vehicle: Experimental Results on ODIN. *IEEE Trans. Control Syst. Technol.* **2001**, *9*, 756–765. [CrossRef]
52. Pettersen, K.Y.; Nijmeijer, H. Underactuated Ship Tracking Control: Theory and Experiments. *Int. J. Control* **2001**, *74*, 1435–1446. [CrossRef]
53. Do, K.D.; Pan, J. Global Tracking Control of Underactuated Ships with Nonzero Off-diagonal Terms in Their System Matrices. *Automatica* **2005**, *41*, 87–95.
54. Asif, M., Khan, M.J., Memon, A.Y. Integral Terminal Sliding Mode Formation Control of Non-Holonomic Robots Using Leader Follower Approach. *Robotica* **2017**, *35*, 1473–1487. [CrossRef]
55. Sun, Z.J.; Zhang, G.Q.; Lu, Y.; Zhang, W.D. Leader-follower Formation Control of Underactuated Surface Vehicles Based on Sliding Mode Control and Parameter Estimation. *ISA Trans.* **2018**, *72*, 15–24. [CrossRef] [PubMed]
56. Fossen, T.I. *Nonlinear Modeling and Control of Underwater Vehicles*; Norwegian Institute of Technology: Trondheim, Norway, 1991.
57. Paliotta, C.; Lefeber, E.; Pettersen, K.Y.; Pinto, J.; Costa, M.; de Figueiredo Borges de Sousa, J.T. Trajectory Tracking and Path Following for Underactuated Marine Vehicles. *IEEE Trans. Control Syst. Technol.* **2019**, *27*, 1423–1437. [CrossRef]
58. Zuo, Z.Y. Nonsingular Fixed-Time Consensus Tracking for SecondOrder Multi-Agent Networks. *Automatica* **2015**, *54*, 305–309. [CrossRef]
59. Zuo, Z.Y.; Han, Q.L.; Ning, B.D.; Ge, X.H.; Zhang, X.M. An Overview of Recent Advances in Fixed-Time Cooperative Control of Multiagent Systems. *IEEE Trans. Ind. Inform.* **2018**, *14*, 2322–2334. [CrossRef]
60. Gao, Z.Y.; Guo, G. Fixed-time Formation Control of AUVs Based on a Disturbance Observer. *ACTA Autom. Sin.* **2019**, *45*, 1094–1102.
61. Gao, J.Q.; Li, Y.J.; Xu, Y.H.; Lv, S.H. A Two-Objective ILP Model of OP-MATSP for the Multi-Robot Task Assignment in an Intelligent Warehouse. *Appl. Sci.* **2022**, *12*, 4843. [CrossRef]
62. Zhao, Y.X.; Wang, M.Y. The LOS/NLOS Classification Method Based on Deep Learning for the UWB Localization System in Coal Mines. *Appl. Sci.* **2022**, *12*, 6484. [CrossRef]
63. Huang, H.H.; Wei, J.Y. In-Motion Coarse Alignment Method Based on Position Loci and Optimal-REQUEST for SINS. *Appl. Sci.* **2022**, *12*, 7113. [CrossRef]

64. Xu, Z.R.; Wang, M.Y.; Li, Q.M.; Qian, L.F. Fault Diagnosis Method Based on Time Series in Autonomous Unmanned System. *Appl. Sci.* **2022**, *12*, 7366. [CrossRef]
65. Sun, H.Y.; Wu, G.Q.; Wang, X.L.; Zhang, T.; Zhang, P.; Chen, W.; Zhu, Q.H. Research on a Measurement Method for the Ocean Wave Field Based on Stereo Vision. *Appl. Sci.* **2022**, *12*, 7447. [CrossRef]
66. Li, Y.S.; Wang, B.; Chen, Y.Q. A Novel Decoupled Synchronous Control Method for Multiple Autonomous Unmanned Linear Systems: Bounded L2-Gain for Coupling Attenuation. *Appl. Sci.* **2022**, *12*, 7551. [CrossRef]
67. Sun, Y.; Cui, B.B.; Ji, F.; Wei, X.H.; Zhu, Y.Y. The Full-Field Path Tracking of Agricultural Machinery Based on PSO-Enhanced Fuzzy Stanley Model. *Appl. Sci.* **2022**, *12*, 7683. [CrossRef]
68. Zhao, M.J.; Zhang, T.; Wang, D. A Novel UWB Positioning Method Based on a Maximum-Correntropy Unscented Kalman Filter. *Appl. Sci.* **2022**, *12*, 12735. [CrossRef]

applied sciences

MDPI

Article

Research on Hard, Transparent and Hydrophobic Coating on PMMA Sheet

Guoqing Wu [1,2,*], Xiaoping Chen [1,2], Xuanyu Xie [3], Pu Zhang [1,2], Shenyu Ge [1], Wei Chen [1], Xian Zeng [3,*] and Ruoye Wang [1]

1 China Ship Scientific Research Center, Wuxi 214000, China
2 Taihu Laboratory of Deepsea Technological Science, Wuxi 214082, China
3 School of Materials Science and Engineering, Wuhan University of Technology, Wuhan 430070, China
* Correspondence: 15852797071@139.com (G.W.); zengxian9925@163.com (X.Z.)

Featured Application: This technology can realize self-cleaning of deep-sea optical windows, reducing the cost of equipment maintenance and cleaning.

Abstract: In this paper, nano SiO_2 particles modified organic silane coatings were successfully prepared to aim at the application of the self-cleaning coating on PMMA substrate for deep-sea optical windows. The chemicals, surface microstructure, wettability, hardness, adhesion, transparency, water scouring resistance as well as microorganism attachment rate of the coatings were investigated. The results showed that adding SiO_2 nanoparticles into the organic silicon coating can effectively improve the hydrophobicity due to generating a micro-nano structure surface. However, excessive addition would result in a decrease in hydrophobicity, adhesion, as well as transparency, due to the inorganic SiO_2 particle destroying the integrity of the organic coating. The optimal coating was obtained by adding 0.5 wt% nano SiO_2 particles, which possessed a water contact angle of 114.2°, hardness of 4H, adhesion level of 0, and visible light transmittance of 0.886. After 40-h water scouring, the water contact angle decreased to 108.3° and the visible light transmittance decreased to 0.839, suggesting good water scouring resistance. The microorganism attachment rate of the S05 coating was 0.17% after a 6 h immersion test, which was about half that of the PMMA substrate.

Keywords: self-cleaning coating; wettability; hardness; adhesion; transparency; water scouring

1. Introduction

Polymethyl methacrylate (PMMA), generally known as plexiglass, is wildly used for optical windows material due to its good light transmission performance, high mechanical strength, good corrosion resistance and easy molding [1]. However, the surface of light-transmitting materials of underwater optical windows will be affected by various factors such as seawater erosion, corrosion and marine biological fouling, which seriously interferes with the performance and stability of underwater optical equipment. By coating self-cleaning coating on the surface of window materials, the cost of equipment maintenance and cleaning can be effectively reduced.

Common self-cleaning coatings are mainly divided into two categories. One is hydrophobic coating with low surface energy, which is mainly composed of organic silane coating [2–4] and organic fluorine coating [5–7], and the other is photocatalytic coating containing TiO_2 [8–10]. It is worth mentioning that the self-cleaning coating applied to the PMMA surface in the above environment needs to meet several characteristics. In addition to good self-cleaning effects, the light transmission, adhesion strength and wear resistance of the coating must also be considered. More importantly, the heat resistance of polymethylmethacrylate is not high. Although its glass transition temperature reaches 104 °C, the maximum continuous use temperature varies from 65 to 95 °C depending on the working conditions, and the thermal deformation temperature is about 96 °C. Therefore, the forming temperature of self-cleaning coating shall not exceed 60 °C, and it is better to have the ability of cure at room temperature.

Contributed to good adhesion, hydrophobicity and corrosion resistance, organic silane coating is an ideal material for preparing self-cleaning coating. Moreover, it also can be formed at room temperature by adjusting the composition of the siloxane monomer [11]. Among the commonly used alkoxysilane monomers, methyltrialkoxysilane and orthosilicate are beneficial to hardness due to high crosslinking, and dimethyldialoxysilane is beneficial to hydrophobicity. In addition, inspired by the lotus leaves, a superhydrophobic surface can be achieved by a combination of low surface energy materials and micro-nanostructures [12–15]. Thus, in this paper, targeting the self-cleaning coating of PMMA optical window, we prepared a hard, transparent and hydrophobic coating by using methyltrialkoxysilane, orthosilicate and dimethyldialoxysilane as the raw material for organic silane coating and nano-silica as a modifier. The chemicals, surface microstructure, wettability, hardness, adhesion, transparency, water scouring resistance as well as microorganism attachment rate of the coatings were systematically investigated in this study, which might provide a new way to solve the problem of deep-see contamination of optical windows.

2. Materials and Methods

2.1. Coatings Preparation

PMMA sheet with a size of 30 mm × 40 mm × 3 mm was used as substrate. Tetraethoxysliane (TEOS), methyltriethoxysilane (MTES), Dimethoxydimethylsilane (DMDS) were used as the precursor of silicon source. Absolute ethanol was used as a solvent. 3–(isobutenyloxy) propyltrimethoxysilane (KH570) was used as a coupling agent. Hydrophobic nano SiO_2 particles (size ranging 7–40 nm) were used as a surface modifier, and the added amounts were 0 (S0), 0.5 wt% (S05), 1 wt% (S1), 2 wt% (S2) correspondingly. Hydrochloric acid (HCl, 37 wt%) was used as a catalyst. The absolute ethanol and hydrochloric acid were purchased from China National Pharmaceutical Group Co., Ltd. (Beijing, China). The rest chemicals were purchased from Shanghai Macklin Biochemical Technology Co., Ltd. (Shanghai, China).

Firstly, the nano SiO_2 particles were ultrasonically dispersed with absolute ethanol for 30 min. Then TEOS, MTES and DMDS with a mole ratio of 3:5:2 were added into the dispersion and mixed for 1 h. After that, a proper amount of deionized water and hydrochloric acid were added to the solution and fully mixed for 1 h at 70 °C, and then cooled to room temperature. Finally, the self-cleaning paint was obtained by adding the coupling agent KH570 and mixing it for 6 h at room temperature. A top pot spay gun was employed to deposit the coating on the PMMA substrate with an air pressure of 28 kPa. The sprayed sample was cured at room temperature for 24 h, pending testing.

2.2. Characterizations

Use a standard contact angle tester (SL-200A, Kono, USA) to measure the wetting angle of the self-cleaning coating, use a UV-VIS-NIR spectrometer (UV3600, Tsushima, Japan) to measure the visible light transmittance of the coating, and use a Fourier transform infrared spectrometer (FTIR, Nicolet6700, Thermophile, USA) to measure the infrared spectrum of the coating. Use a cold field emission scanning electron microscope (SEM, JSM-7500F, Japan Electronics Co., Ltd.) to observe the microstructure of the coating surface, use a hand-pushed pencil hardness tester (QHQ-A, Zhejiang Aipu) to measure the hardness of the coating according to ISO 15184:2020 Paints and Varnishes-Determination of Film Hardness by pencil test, and use a Baige knife (QFH-A, Zhejiang Airuipu) to measure the adhesion of the coating according to ISO 2409:2020 Paints and Varnishes-Cross-Cut Test. The judgment standards of coating adhesion are shown in Table 1.

Table 1. Classification of test results for coating adhesion.

Classification	Description
0	The edges of the cuts are completely smooth; none of the squares of the lattice is detached.
1	Detachment of small flakes of the coating at the intersections of the cuts. A cross-cut area not greater than 5% is affected.
2	The coating has flaked along the edges and/or at the intersections of the cuts. A cross-cut area greater than 5%, but no greater than 15%, is affected.
3	The coating has flaked along the edges of the cuts partly or wholly in large ribbons, and/or it has flaked partly and wholly on different parts of the squares. A cross-cut area greater than 15%, but not greater than 35%, is affected.
4	The coating has flaked along the edges of the cuts in large ribbons and/or some squares have detached partly or wholly. A cross-cut area greater than 35%, but not greater than 65%, is affected.
5	Any degree of flaking that cannot even be classified by classification 4.

The water-flow scouring resistance of the prepared coating was evaluated through a laboratory self-made test device as shown in Figure 1, where the rotor stirred the deionized water to form an annular flow. The coating sample was fixed at the same height level as the rotor blades, and the coating surface was perpendicular to the radial direction. The rotor radius R was 40 mm, and the distance d from the coating surface to the axis was 50 mm. The rotor speed was 2000 rad/min. The wettability and transparency after 20 h and 40 h scouring were detected.

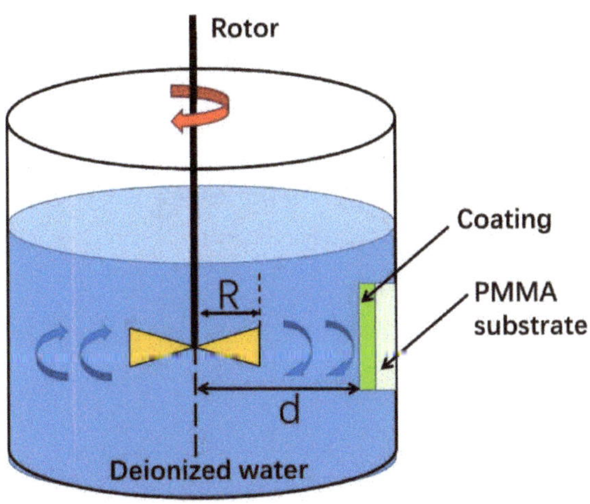

Figure 1. Schematic diagram of water-flow scouring test.

The microorganism attachment behavior of the prepared coating was evaluated by a microbial immersion test. Specifically, immerse the sample in sterilized seawater for 12 h, and then take it out and immerse it in the phaeodactylum tricornutum solution with a concentration of 2×10^6 for 6 h. After the immersion test, the sample surface is washed with ultrapure water for 30 s. The microorganism attachment rate is calculated by counting the adhesion area rate with a fluorescence microscope (FM, ZEISS, Germany). A total of 20 random fields magnified by 40 times were taken by the fluorescence microscope for each sample. The adhesion area was evaluated via an Image processing software (Image pro-plus).

3. Results and Discussions

3.1. Chemicals of Coatings

In order to identify the chemicals of prepared coatings, infrared spectral analysis was performed and the results are shown in Figure 2. It was seen that all the coatings exhibited similar infrared spectrums. The main absorption peaks located at 3450 cm^{-1} and 1109 cm^{-1} belonged to the vibrations of Si-OH and Si-O-Si correspondingly, indicating most siloxanes had been hydrolyzed and polycondensation. The peak at 2969 cm^{-1}, 1260 cm^{-1} and 849 cm^{-1} belonged to the stretching vibrations of Si-O-C_2H_5 [16], suggesting insufficient hydrolysis of siloxane monomer. Specifically, the peak at 2969 cm^{-1} was related to the C-H stretching vibration of residual CH_3 from non-hydrolyzed Si-O-C_2H_5 groups, and the small peaks near and below 2969 cm^{-1} were assigned to asymmetric and symmetric C-H vibrations of CH_2. Moreover, the peaks between 1500–1260 cm^{-1} were assigned to various C-H vibrations, and around 849 cm^{-1} with a band of 900–750 cm^{-1} were assigned to CH_2 [17] Due to the addition of KH570, the vibration absorption peaks of olefin stretching (1702 cm^{-1}) and C=O ester contraction (1723 cm^{-1}) were also observed [18]. The peaks located in 400–500 cm^{-1} belonged to the absorption peaks of Si-O-Si [19].

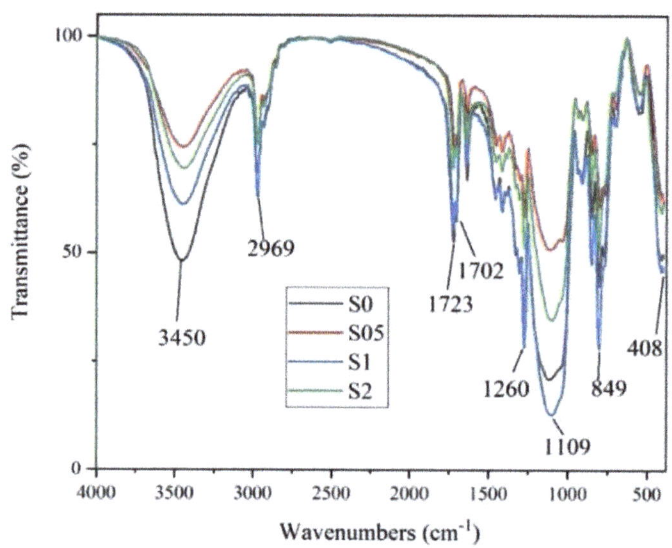

Figure 2. Infrared spectra of nano SiO_2 particles modified coatings with different weight ratio.

In an acid environment, the alkoxysilane monomers would undergo hydrolysis and condensation reactions according to Figure 3. Organic groups (methyl, phenyl, etc.) directly linked to silicon atoms do not participate in the hydrolysis and condensation reactions, while the main one involved in the reaction process is oxyethyl and oxymethyl. When using acid as a catalyst, H^+ attacks -OCH_2CH_3 and/or -OCH_3 groups and results in the protonation and the shift of extranuclear electron cloud to it, weakening the Si-O bond until it breaks and forms Si-OH. So far, one-step hydrolysis has been completed. After hydrolysis, the generated Si-OH will further participate in the condensation reaction with Si-OH, Si-OCH_2CH_3 and Si-OCH_3 under the catalyst action of acid, forming a three-dimensional network structure of Si-O-Si, water and alcohol substance. It is worth noting that although a single monomer is hydrolyzed first and then condensed, hydrolysis and condensation occur simultaneously in the whole system, so incomplete hydrolysis may occur.

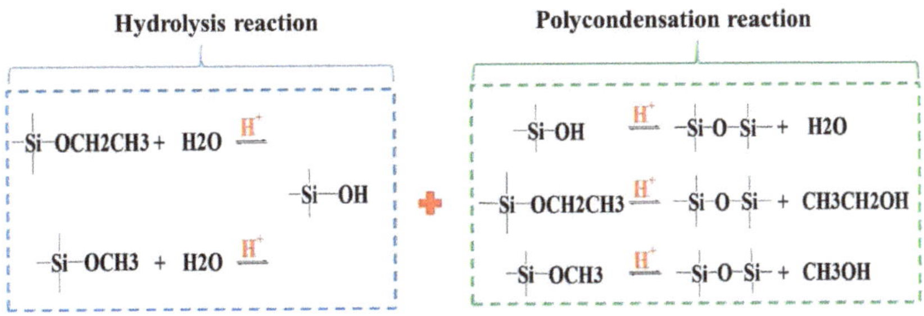

Figure 3. Schematic diagram of hydrolysis and condensation reactions of alkoxysilane monomers.

3.2. Surface Morphology of Coatings

The surface morphology of prepared coatings is shown in Figure 4. It was found that the surface of organic silicon coating without nano SiO_2 particles was relatively smooth and homogeneous. However, the surface of coatings after modification became rougher and rougher as increasing the amount of nano SiO_2 particles. The micro-bulge structure was formed by introducing nanoparticles, which increased the surface roughness apparently. Furthermore, no cracks or pores were observed on the surface, indicating good compatibility between nanopowders and base materials. However, the agglomeration of nano SiO_2 particles occurred properly, owing to the insufficient dispersion of SiO_2, where the sizes of those micro-bulge structures were much larger than that of a single nanoparticle.

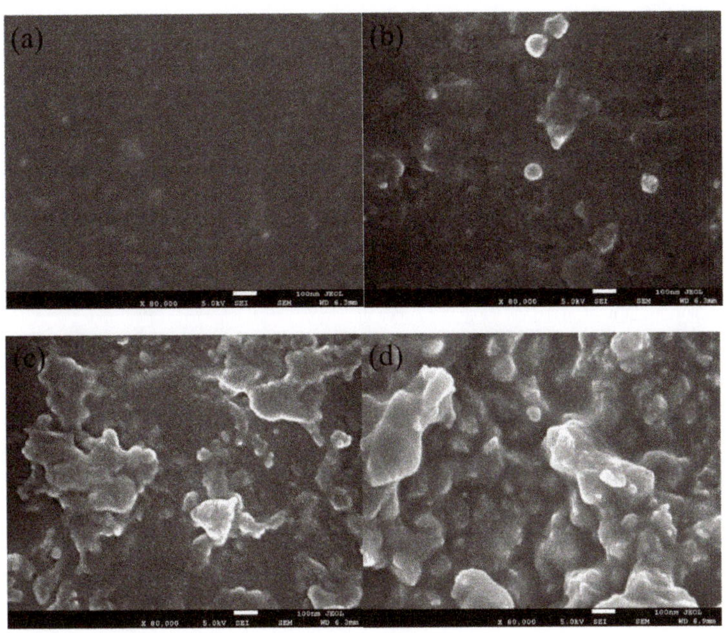

Figure 4. Surface morphology of nano SiO_2 particles modified coatings with different weight ratio (**a**) 0, (**b**) 0.5 wt%, (**c**) 1 wt%, and (**d**) 2 wt%.

3.3. Wettability of Coatings

The wettability of the coating was evaluated by testing the average water contact angle of five samples. The wettability of coatings against nano SiO_2 content is shown in Figure 5. According to it, the average water contact angle of bare PMMA substrate was 79.3°, while after coating, the surface water contact angle was improved above 100°

showing hydrophobic properties. The greatest average water contact angle was 114.2° when the coating was modified with 0.5 wt% nano SiO_2 particles. However, the water contact angle decreased when the amount of SiO_2 continued to increase. It is generally accepted that surface micro-nano structure improves the hydrophobicity, which works similarly to lotus leaves. On the one hand, this contributed to the low surface energy of silicone coating and the micro-nano structure induced by nano SiO_2 particles, the surface water contact angle of PMMA increased after coating and showed hydrophobic properties, which can reduce the adhesion of marine microorganisms or make the pollutants easy to clean by seawater. On the other hand, due to the physical absorption of water on the SiO_2 particle surface [20], the water contact angle would decrease when massive nano SiO_2 particles were added.

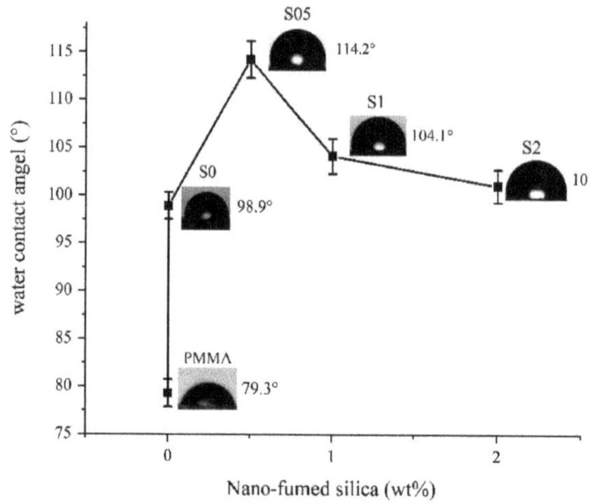

Figure 5. Water contact angles of samples with different nano SiO_2 particles.

3.4. Hardness and Adhesion of Coatings

The hardness of PMMA substrate and coatings was tested according to ISO 15184:2020 Paints and Varnishes-Determination of film hardness by pencil test. The pencil is tilted at 45° to the coating by the hand-pushed pencil hardness tester, and the load applied on the coating surface by the tip of the pencil is 750 g. The cross-cut method was used to detect the adhesion of coatings according to the ISO 2409:2020 Paints and Varnishes-Cross-Cut test. During the test, six parallel cuts are introduced in the coating by the multi-blade cutting tool, and then another six cuts are introduced perpendicular to the first cuts. The spacing of cuts was set as 1 mm. before qualifying the adhesion levels, the loose particles of coatings were removed by brushing and using pressure-sensitive adhesive tape. The detected results of coating hardness and adhesion are listed in Table 2.

Table 2. Hardness and adhesion levels of PMMA and coatings.

Samples	Hardness	Adhesion *Levels*
PMMA	*B*	-
S0	*4H*	0
S05	*4H*	0
S1	*4H*	0
S2	*4H*	1

As shown, the hardness of the coating was higher than the that of PMMA substrate, suggesting better wear resistance of the coating samples. In addition, the added amounts of nano SiO_2 particles were too low to have effects on the coating hardness, which showed the same hardness of 4H. However, the adhesion of the S2 coating was worse than the other coatings, showing an adhesion level of 1. The continuity and integrity of the organic coating would be destroyed by adding inorganic SiO_2 particles, generating more interfaces, which might develop into crack sources, thus reducing the adhesion of the coating.

3.5. Transparency of Coatings

The spectrum transmittance of PMMA and coatings modified with different nano SiO_2 particle weight ratios in the wavelengths of 380 nm–800 nm was detected and shown in Figure 6. It was found that the spectrum transmittance of the coatings was lower than the substrate in the whole band. In addition, the transmittance declined with the increasing weight ratio of SiO_2 due to the scattering of nanoparticles. The total transmittance of samples in 380 nm~800 nm was calculated according to the following equation based on JISR 3106-2019.

$$\tau_v = \frac{\sum_\lambda D\lambda \cdot V\lambda \cdot \tau(\lambda)}{\sum_\lambda D\lambda \cdot V\lambda}$$

where τ_v is visible light transmittance, $D\lambda$ is the relative spectral energy distribution of CIE daylight D_{65}, $V\lambda$ is CIE light adaptation relative luminous efficiency, $\tau(\lambda)$ is the spectrum transmittance. The calculated results are listed in Table 3. It was found that the lowest visible light transmittance 0.835 was obtained with the S2 coating, which was only decreased by 7.3% as compared to that of bare PMMA substrate, suggesting good transparency of the prepared coatings.

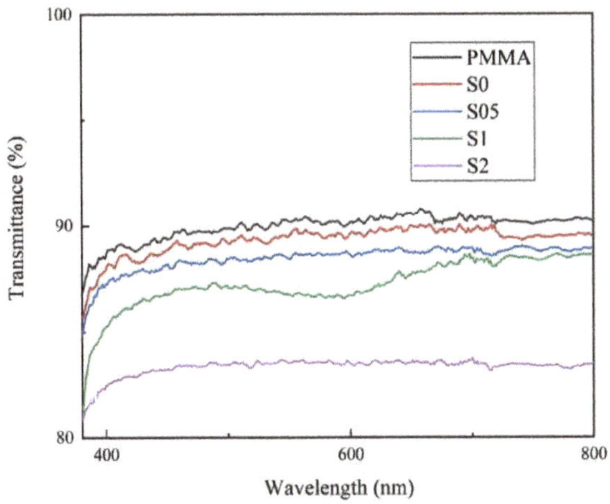

Figure 6. Visible light transmission spectra of PMMA and nano SiO_2 particles modified coatings.

Table 3. Visible light transmittance of PMMA and coatings.

Samples	PMMA	S0	S05	S1	S2
τ_v (%)	90.1	89.5	88.6	86.9	83.5

3.6. Scouring Resistance of Coatings

Relating to the wettability, hardness, adhesive strength and transparency of the coatings comprehensively, the S05 coating was chosen as the optimal coating and underwent the water scouring test. As shown in Figure 7, the wettability of the coating slightly increased, where the water contact angles were 110.6° after 20 h and 108.3° after 40 h. The visible light transmission spectra of the S05 coating after the scouring test are shown in Figure 8, implying that the spectral transmittance was decreased in the whole observed wave band. The total visible light transmittance of the coating was 85.6% after 20 h and

83.9% after 40 h, correspondingly. The roughness of the coating surface increased with increasing the scouring time, which reduced the transparency of the coating. Meanwhile, the coating surface area increased, which promoted the wetting behavior between the water drop and the coating.

Figure 7. Water contact angles of S05 coatings with different scouring time (**a**) 0 h, (**b**) 20 h, and (**c**) 40 h.

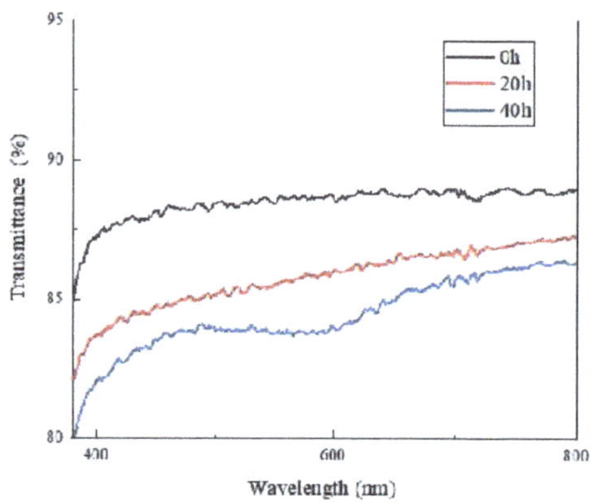

Figure 8. Visible light transmission spectra of S05 coatings with different scouring time.

3.7. Microrganism Attachment Rate of Coatings

The microorganism attachment rates of the PMMA substrate and the S05 coating were measured by counting the mean attachment area ratio of 20 random fields magnified by 40 times. As shown in Figure 9, the attachment area (indexed by pink color) of phaeodactylum tricornutum on the PMMA substrate was apparently larger than that of the S05 coating. Attributed to a better hydrophobicity, the attachment of phaeodactylum tricornutum on the coating surface was weaker and easier to wash away by the water flow. The calculated microorganism attachment rate of the S05 coating was 0.17%, which was about half that of the PMMA substrate.

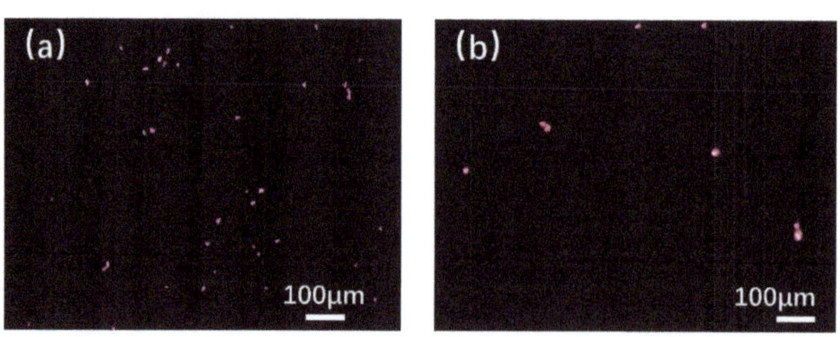

Figure 9. FM images of samples after microorganism immersion test for (**a**) PMMA sheet and (**b**) S05 coating.

4. Conclusions

In this paper, targeting the application of a self-cleaning coating on PMMA substrate for deep-sea optical windows, nano SiO_2 particles modified organic silane coatings were successfully prepared. The chemicals, surface microstructure, wettability, hardness, adhesion, transparency, water scouring resistance as well as microorganism attachment behavior of the coatings were investigated. The results showed that adding SiO_2 nanoparticles into the organic silicon coating can effectively improve the hydrophobicity due to generating a micro-nano structure surface. However, excessive addition would result in a decrease in hydrophobicity, adhesion as well as transparency, due to the inorganic SiO_2 particle destroying the integrity of the organic coating. The optimal coating was the S05 coating, which possessed a water contact angle of 114.2°, hardness of 4H, adhesion level of 0, and visible light transmittance of 0.886. After 40-h water scouring, both the water contact angle and the visible light transmittance were slightly decreased, which were 108.3° and 0.839 respectively, implying good water scouring resistance. Moreover, the microorganism attachment rate of the S05 coating was about half that of the PMMA substrate, suggesting good application prospects for deep-sea optical windows.

Author Contributions: Conceptualization, G.W.; methodology, X.X. and X.C.; formal analysis, P.Z., S.G. and W.C.; Writing—review and editing, X.Z.; Project administration, R.W. All authors have read and agreed to the published version of the manuscript.

Funding: This research was funded by the Early-Stage Key Technology Research Project of Cold Seep Equipment from the Chinese Academy of Science (LQ-GJ-01).

Institutional Review Board Statement: Not applicable.

Informed Consent Statement: Not applicable.

Data Availability Statement: The data that were used to support the findings of this study are included within the article.

Conflicts of Interest: The authors declare no conflict of interest.

References

1. Xia, F.; Yang, L.; Dai, B.; Gao, G.; Yang, Z.; Cao, W.; Xu, L. Novel reparation method for polymethyl methacrylate optical windows of aircrafts damaged by service environment. *Sci. China Technol. Sci.* **2016**, *101*, 385–390. [CrossRef]
2. Matin, A.; Merah, N.; Ibrahim, A. Superhydrophobic and self-cleaning surfaces prepared from a commercial silane using a single-step drop-coating method. *Prog. Org. Coat.* **2016**, *99*, 322–329. [CrossRef]
3. Adark, D.; Bhattacharyya, R.; Saha, H.; Maiti, P.C. Sol gel processed silica based highly transparent self-cleaning coatings for solar glass covers. *Mater. Today Proc.* **2020**, *33*, 2429–2433. [CrossRef]
4. Kumar, D.; Wu, X.; Fu, Q.; Weng, J.; Ho, C.; Kanhere, P.D.; Li, L.; Chen, Z. Development of durable self-cleaning coatings using organic-inorganic hybrid sol-gel method. *Appl. Surf. Sci.* **2015**, *344*, 205–212. [CrossRef]
5. Cui, Y.; Wei, B.; Wang, Y.; Guo, X.; Xiao, J.; Li, W.; Pang, A.; Bai, Y. Fabrication of UV/moisture dual curing coatings based on fluorinated polyoxetanes for anti-fouling applications. *Prog. Org. Coat.* **2022**, *163*, 106656. [CrossRef]
6. Yu, F.; Gao, J.; Liu, C.; Chen, Y.; Zhong, G.; Hodges, C.; Chen, M.; Zhang, H. Preparation and UV aging of nano-SiO2/fluorinated polyacrylate polyurethane hydrophobic composite coating. *Prog. Org. Coat.* **2020**, *141*, 10556. [CrossRef]
7. Kumar, D.; Li, L.; Chen, Z. Mechanically robust polyvinylidene fluoride (PVDF) based superhydrophobic coatings for self-cleaning applications. *Prog. Org. Coat.* **2016**, *101*, 385–390. [CrossRef]
8. Xu, F.; Wang, T.; Chen, H.; Bohling, J.; Maurice, A.M.; Wu, L.; Zhou, S. Preparation of photocatalytic TiO2-based self-cleaning coatings for painted surface without interlayer. *Prog. Org. Coat.* **2017**, *113*, 15–24. [CrossRef]
9. Chen, P.; Wei, B.; Zhu, X.; Gao, D.; Gao, Y.; Cheng, J.; Liu, Y. Fabrication and characterization of highly hydrophobic rutile TiO2-based coatings for self-cleaning. *Ceram. Int.* **2019**, *45*, 6111–6118. [CrossRef]
10. Zong, L.; Wu, Y.; Li, X.; Jiang, B. The preparation of superhydrophobic photocatalytic fluorosilicone/SiO2–TiO2 coating and its self-cleaning performance. *J. Coat. Technol. Res.* **2021**, *18*, 1245–1259. [CrossRef]
11. Zhan, Y.; Grottenmüller, R.; Li, W.; Javaid, F.; Riedel, R. Evaluation of mechanical properties and hydrophobicity of room-temperature, moisture-curable polysilazane coatings. *J. Appl. Polym. Sci.* **2021**, *138*, 50469. [CrossRef]
12. Jeevahan, J.; Chandrasekaran, M.; Britto Joseph, G.; Durairaj, R.B.; Mageshwaran, G. Superhydrophobic surfaces: A review on fundamentals, applications, and challenges. *J. Coat. Technol. Res.* **2018**, *15*, 231–250. [CrossRef]
13. Bhushan, B.; Multanen, V. Designing liquid repellent, icephobic and self-cleaning surfaces with high mechanical and chemical durability. *Philos. Trans. R. Soc. A Math. Phys. Eng. Sci.* **2019**, *377*, 20180270. [CrossRef] [PubMed]

14. Hooda, A.; Goyat, M.S.; Pandey, J.K.; Kumar, A.; Gupta, R. A review on fundamentals, constraints and fabrication techniques of superhydrophobic coatings. *Prog. Org. Coat.* **2020**, *142*, 105557. [CrossRef]
15. Farzaneh, M. Ice accretions on high-voltage conductors and insulators and related phenomena. *Philos. Trans. R. Soc. A Math. Phys. Eng. Sci.* **2000**, *358*, 2971–3005. [CrossRef]
16. Hu, C.; Xu, G.; Shen, X.; Shao, C.; Yan, X. The epoxy-siloxane/Al composite coatings with low infrared emissivity for high temperature applications. *Appl. Surf. Sci.* **2010**, *256*, 3459–3463. [CrossRef]
17. Öhman, M.; Persson, D. ATR-FTIR Kretschmann spectroscopy for interfacial studies of a hidden aluminum surface coated with a silane film and epoxy I. Characterization by IRRAS and ATR-FTIR. *Surf. Interface Anal.* **2012**, *44*, 133–143. [CrossRef]
18. Peng, C.; Wu, Z.; Li, J.; Wang, Z.; Wang, H.; Zhao, M. Synthesis, thermal and mechanical behavior of a silicon/phosphorus containing epoxy resin. *Appl. Polym.* **2015**, *132*, 1–10. [CrossRef]
19. Du, Y.; Wang, C.; Yang, L.; Zhu, S.; Wang, F. Enhanced oxidation and corrosion inhibition of 1Cr11Ni2W2MoV stainless steel by nano-modified silicone-based composite coatings at 600 °C. *Corros. Sci.* **2020**, *169*, 108599. [CrossRef]
20. Machia, M.; Norimoto, K.; Watanabe, T.; Hashimoto, K.; Fujishima, A. The effect of SiO_2 addition in super-hydrophilic property of TiO_2 photocatalyst. *J. Mater. Sci.* **1999**, *34*, 2569–2574. [CrossRef]

applied sciences

MDPI

Article

A Novel UWB Positioning Method Based on a Maximum-Correntropy Unscented Kalman Filter

Mujie Zhao [1,*], Tao Zhang [1,*] and Di Wang [2]

1 Key Laboratory of Micro-Inertial Instrument and Advanced Navigation Technology, Ministry of Education, School of Instrument Science and Engineering, Southeast University, Nanjing 211100, China
2 College of Energy and Electrical Engineering, Hohai University, Nanjing 211100, China
* Correspondence: zhaomujie168@163.com (M.Z.); zhangtao22@seu.edu.cn (T.Z.)

Abstract: Aiming at the problem of measurement-information abnormal-error and nonlinear filtering in UWB navigation and positioning, an ultra wideband position algorithm based on a maximum cross-correlation entropy unscented Kalman filter is proposed. The algorithm first obtains the predictive state estimate and the covariance matrix through traceless transformation. Then, it reconstructs observation information using the nonlinear regression method based on the maximum cross-correlation entropy criterion, which enhances the robustness of the unscented Kalman filter algorithm for heavy-tailed noise. The simulation and actual test results show that this algorithm has better positioning accuracy and stability than the traditional filter algorithm in a non Gaussian noise environment. This algorithm effectively solves the problem that UWB indoor location is easily affected by indoor environments, resulting in fixed deviation for that location.

Keywords: UWB; UKF; position; maximum correntropy

1. Introduction

In recent years, with the advent of the Internet of Things era, all aspects of human life have basically been related to location-based services. For example, in large supermarkets, people can quickly and conveniently find particular goods. In large office buildings, people can find particular locations accurately and quickly. Parents can discern their children's location in real time to avoid losing them. Vehicles in a parking lot can automatically find parking spaces and park. These services are inseparable from accurate location information. Positioning and navigation technology has become the key to smart city construction, which directly affects people's safety and economic development. Therefore, research and implementation of location technology based on wireless technology has an extremely high engineering-application value [1].

The global navigation satellite system (GNSS) is a positioning system based on satellites; it uses satellites as basic reference objects. Because of its high positioning accuracy and its high strategic value, GNSS has now become a focus of all countries and has independently developed its own navigation systems. Compared with other positioning systems, it has unique advantages. Satellite signals can be sent to the global scope, enabling people to obtain positioning information anytime and anywhere. In addition, the satellite system has other characteristics, such as high positioning accuracy, high robustness and all-weather free-space optical communication technology. However, because transmission of satellite signals over a long distance is very weak by the time the signals reach the ground, those signals could rapidly decline or even be unable to be used under the influence of buildings on the ground [2]. Generally, satellite navigation system is suitable for open outdoor environments, while in indoor environments, people urgently need a high-precision, highly reliable indoor positioning system to meet their growing needs.

Ultra wideband (UWB) technology is a new type of wireless communication technology. At this stage, it is mainly used to obtain point-to-point distance based on time and achieve final positioning with the distances between the label to be tested and multiple reference base stations. Because UWB technology uses narrow nanosecond nonsinusoidal pulses for data transmission and its spectrum range is very wide, in theory, UWB technology can achieve centimeter-level positioning accuracy [3]. In addition, UWB technology also has the advantages of low transmission power, strong multipath anti-interference capability and high security, so it is very suitable for indoor and other closed-area positioning applications.

Non line of sight (NLOS) environments affect the positioning accuracy of UWB technology, and many scholars have put forward their own opinions on reducing NLOS error. A study in the Reference [4], proposed a tight combination of UWB/inertial navigation system (INS) indoor positioning and an attitude-determination method. The difference between the round trip time (RTT) ranging information corrected with a standard time deviation and the distance information calculated with an INS was used as measurement information. NLOS errors were eliminated according to the set threshold, and indoor position and attitude determination were conducted through an extended Kalman filter. Reference [5] proposes an adaptive robust Kalman filtering method that uses the threshold to construct a robust factor that identifies and weakens NLOS ranging errors. Meanwhile, the same study used a Sage Husa filter to estimate and correct system noise in real time in order to improve the accuracy of the UWB positioning. Another study in the Reference [6], analyzed the characteristics of the clock that is offset of the UWB antenna in the actual environment, the relative speed between the nodes and the ranging errors caused by the non line of sight environment. Meanwhile, the same study constructed an error-compensation method to achieve positioning and improved the ranging and positioning accuracy of UWB positioning in practical applications. A third study in the literature, [7], estimated location information based on Chan and Taylor's collaborative location method via setting a threshold value for the residual of estimated results to identify the NLOS environment, calculating location results using the Kalman method for qualified measurement data, weighting the residual and moving averages of the location results and completing an update of the final location. Reference [8] proposed a robust-volume Kalman filter (CKF) algorithm with a noise–time-varying estimator. The robust equivalent covariance matrix was constructed via use of the predicted residual factor to control the influence of observation outliers on the filter-parameter solution. The Sage Husa algorithm was used to estimate and correct the statistical characteristics of system noise in real time, improve filtering accuracy and stability and achieve high-precision positioning. Reference [9] proposed a robust Student's t-based Kalman filter, which provided a Gaussian approximation of posterior distribution.

In order to avoid deviation of UWB position estimation caused by linearization, some researchers used an unscented Kalman filter (UKF) for UWB position estimation, estimated the position through traceless transformation of sigma points and obtained the UWB position estimation algorithm based on the UKF [10,11]. However, the algorithms based on the extended Kalman filter (EKF) and the UKF both assume that the state noise and the measurement noise will obey Gaussian distribution; therefore, they are not suitable for non Gaussian noise. However, in the UWB indoor-positioning process, non Gaussian noise, such as the noise of mobile cars, pulse noise caused by electromagnetic interference and communication system faults or defects, does exist objectively [12,13].

Aiming at the problem of measurement-information abnormal-error and nonlinear filtering for UWB positioning, this paper proposes a maximum cross-correlation entropy-based unscented Kalman filter algorithm based on the maximum cross-correlation criterion on the basis of a nonlinear filter algorithm. The algorithm first obtains a predictive state estimate and a covariance matrix through traceless transformation and then reconstructs observation information using the nonlinear regression method based on the maximum cross-correlation entropy criterion, which enhances robustness of the unscented Kalman filter algorithm.

This paper is arranged as follows: Section 2 introduces the UWB positioning method based on nonlinear filtering, including the principle of UWB positioning and the UWB positioning method based on the unscented Kalman filter. Section 3 introduces the proposed maximum cross-correlation entropy UKF algorithm, including the maximum cross-correlation entropy criterion and the proposed improved UKF algorithm. In Section 4, the effectiveness of the proposed method is demonstrated through simulation and an experiment. The fifth section is the conclusion.

2. UWB Positioning Method Based on Unscented Kalman Filter

2.1. Principle of UWB Positioning

At present, common range-based wireless positioning methods include the signal-arrival time-based method, the signal-arrival time-difference-based method, the signal-arrival angle-based method and the received-signal strength-based method [14–16]. UWB technology is carrier-free communication technology. It uses narrow nanosecond nonsinusoidal pulses to transmit data. It can measure precise signal transmission time, which can accurately measure the distance between the base station and the label. Therefore, UWB technology usually adopts a time of arrival (TOA)-based localization algorithm.

The TOA positioning method is also called the circle positioning method, which calculates the coordinates of a position label by finding the intersection points of circles [17]. The location of the base station (BS) is used as the center of each circle, and the distance from the BS to the location label is used as the radius to draw each circle. If there is no error in the distances, the final circles will intersect at a point, which is the label position.

Based on the TOA model, a schematic diagram of the positioning principles of three base stations and one label is established on a plane. The base-station coordinate is defined as $\mathrm{BS}(x_i, y_i)$, and the coordinates of the positioning label are Label (x, y). The measured distance from the base station to the label is R_i. The distance with the base station as the center, R_i is the radius, which forms the circle-position line. Due to the existence of ranging errors, the positioning model is shown in Figure 1.

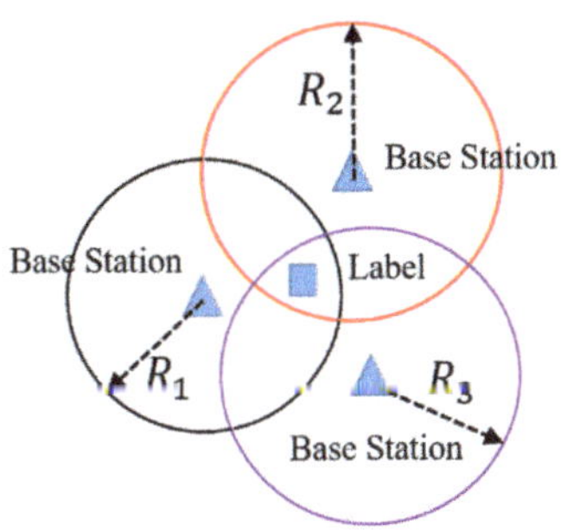

Figure 1. TOA-based geometric positioning model.

Since the coordinates, $\mathrm{BS}(x_i, y_i)$, of the three base stations and the distance between the three base stations and the label are known, the relationship between the coordinate position and the distance can be established. Through the analysis above, the following equations can be obtained:

$$\begin{cases} (x_1 - x)^2 + (y_1 - y)^2 = R_1^2 \\ (x_2 - x)^2 + (y_2 - y)^2 = R_2^2 \\ (x_3 - x)^2 + (y_3 - y)^2 = R_3^2 \end{cases} \tag{1}$$

Calculate the difference between two pairs of the equations above to formulate

$$\begin{cases} R_2^2 - R_1^2 = \left(x_2^2 + y_2^2\right) - \left(x_1^2 + y_1^2\right) - 2(x_2 - x_1)x - 2(y_2 - y_1)y \\ R_3^2 - R_1^2 = \left(x_3^2 + y_3^2\right) - \left(x_1^2 + y_1^2\right) - 2(x_3 - x_1)x - 2(y_3 - y_1)y \end{cases} \tag{2}$$

The formula above is written in matrix form as follows:

$$\begin{bmatrix} x_2 - x_1 & y_2 - y_1 \\ x_3 - x_1 & y_3 - y_1 \end{bmatrix} \begin{bmatrix} x \\ y \end{bmatrix} = \frac{1}{2} \begin{bmatrix} \left(x_2^2 + y_2^2\right)^2 - \left(x_1^2 + y_1^2\right) + R_1^2 - R_2^2 \\ \left(x_3^2 + y_3^2\right)^2 - \left(x_1^2 + y_1^2\right) + R_1^2 - R_3^2 \end{bmatrix} \tag{3}$$

The coordinates of the label can be obtained from the formula above.

$$\begin{bmatrix} x \\ y \end{bmatrix} = \frac{1}{2} \begin{bmatrix} x_2 - x_1 & y_2 - y_1 \\ x_3 - x_1 & y_3 - y_1 \end{bmatrix}^{-1} \begin{bmatrix} \left(x_2^2 + y_2^2\right)^2 - \left(x_1^2 + y_1^2\right) + R_1^2 - R_2^2 \\ \left(x_3^2 + y_3^2\right)^2 - \left(x_1^2 + y_1^2\right) + R_1^2 - R_3^2 \end{bmatrix} \tag{4}$$

The position of the label can be obtained through solving Equation (4). However, in most cases, there are more than three base stations. Therefore, the least squares method (LSM) is usually used to solve Equation (4) so as to obtain a more accurate label position. The TOA method requires high clock synchronization between the labels and the BS. If the clock is not synchronized, the signal time will be incorrect when the label arrives at the BTS. If erroneous data is used for positioning, the positioning accuracy of the positioning system will decline, so in order to achieve high-precision positioning, the clock synchronization between the label and the BTS must be guaranteed, though that is difficult to achieve in practice.

2.2. UWB Location Algorithm Based on the UKF

Formula (4) obtains the initial positioning coordinates of unknown nodes. The UKF algorithm is modified according to the characteristics of the NLOS environment to perform precise positioning and position tracking. The UKF is a new nonlinear filter estimation algorithm that is UT (unscented)-based. The UKF and the EKF have different approaches to linearizing nonlinear functions. The UKF uses UT transformation to deal with nonlinear transference of the mean and the covariance, and directly uses the nonlinear model of the system. The EKF approximates nonlinear functions and needs the Jacobian matrix to be calculated, which greatly increases the amount of calculation. The UKF algorithm obtains more observational assumptions through generation of Sigma points and ignores higher-order terms without linearization. Therefore, in terms of nonlinear problems in an NLOS environment, the UKF has higher calculation accuracy and stronger adaptability than has the EKF [18].

The state equation is established as follows:

$$\dot{X} = FX + W \tag{5}$$

where F is the state transition matrix and W represents the matrix of noise. X represents the state vector, which can be expressed as

$$X = \left[P_x\ P_y\ V_x\ V_y\right]^T \tag{6}$$

where P_x and P_y represent the positioning coordinates of the label in the X direction and the Y direction, respectively. V_x and V_y indicate the motion velocity of the label in the X and Y directions, respectively. F from Equation (5) is

$$F = \begin{bmatrix} 1\,0\,\Delta T\,0 \\ 0\,1\,0\,\Delta T \\ 0\,0\,1\,0 \\ 0\,0\,0\,1 \end{bmatrix} \tag{7}$$

According to the matrix of noise, the covariance matrix, Q, is

$$Q = q\begin{bmatrix} \Delta T^3/3\ 0\ \Delta T^2/2\ 0 \\ 0\ \Delta T^3/3\ 0\ \Delta T^2/2 \\ \Delta T^2/2\ 0\ \Delta T\ 0 \\ 0\ \Delta T^2/2\ 0\ \Delta T \end{bmatrix} \tag{8}$$

where q is the power spectrum density of the system noise. ΔT is the UWB-data sampling interval. According to the relationship between the position and the distance of the two coordinate points, the equation for the distance between the label and the base station can be established. Considering the existence of multiple base stations, the following general equation is given:

$$R_{i,k} = \sqrt{(x_k - x_i)^2 + (y_k - y_i)^2} + n_{i,k} \tag{9}$$

where $R_{i,k}$ represents the measurement distance and $n_{i,k}$ is measurement noise. When the viewing-distance environment is between the base station and the mobile station, $n_{i,k}$ follows zero-mean Gaussian distribution. When the base station and mobile station are in a non line of sight environment or abnormal ranging occurs, then the measurement equation is

$$Z = HX + V \tag{10}$$

where H represents the measurement transfer matrix and V represents measurement noise. Measurement matrix Z consists of the distance measured with UWB positioning:

$$Z = [R_{1,k}\ R_{2,k}\ \cdots\ R_{M,k}]^T \tag{11}$$

$$H = \begin{bmatrix} \sqrt{(x_k - x_1)^2 + (y_k - y_1)^2} \\ \sqrt{(x_k - x_2)^2 + (y_k - y_2)^2} \\ \cdots \\ \sqrt{(x_k - x_M)^2 + (y_k - y_M)^2} \end{bmatrix} \tag{12}$$

$$V = [n_{1,k}\ n_{2,k}\ \cdots\ n_{M,k}]^T \tag{13}$$

To sum up, the steps of the UWB positioning algorithm based on the UKF are as follows:
Step 1: Set the initial parameters of the UKF filter:

$$\begin{cases} \hat{X}_0 = E(\hat{X}_0) \\ P_0 = E\left[(X - X_0)(X - X_0)^T\right] \end{cases} \tag{14}$$

Step 2: According to Equation (14), calculate the sigma sample points:

$$\chi_{k-1} = \left[\hat{X}_{k-1}\hat{X}_{k-1} + \gamma\sqrt{P_{k-1}}\hat{X}_{k-1} - \gamma\sqrt{P_{k-1}}\right]^T \tag{15}$$

Among them, P_{k-1} represents the covariance matrix, χ_{k-1} represents the composed column vector and γ represents a scale factor.

Step 3: Calculate the covariance matrix and the predicted state vector:

$$\begin{cases} \chi_{k|k-1} = f(\chi_{k-1}) \\ \hat{X}_{k|k-1} = \Sigma_{i=0}^{2n} w_i^m \chi_{k|k-1}^i \\ P_{k|k-1} = \Sigma_{i=0}^{2n} w_i^c \left[\chi_{k|k-1}^i - \hat{X}_{k|k-1}\right] \end{cases} \tag{16}$$

Among these variables, w_i^m is the weight of the mean value and w_i^c is the weight of the covariance.

Step 4: With $\hat{X}_{k|k-1}$ and P_{k-1} as the output values, calculate the sigma sample points according to Step 2 (ζ_{k-1}), then calculate the measurement information predicted at time k:

$$\zeta_{k-1} = \left[\hat{X}_{k|k-1}\hat{X}_{k|k-1} + \gamma\sqrt{P_{k|k-1}}\hat{X}_{k|k-1} - \gamma\sqrt{P_{k|k-1}}\right]^T \tag{17}$$

Step 5: Measure and update to obtain the covariance matrix and the predicted state vector:

$$\begin{cases} \zeta_{k|k-1} = h(\zeta_{k-1}) \\ \hat{Z}_{k|k-1} = \sum_{i=0}^{2n} w_i^m \zeta_{k|k-1}^i \end{cases} \tag{18}$$

$$P_{xz} = \sum_{i=0}^{2n} w_i^c \left[\chi_{k|k-1}^i - \hat{X}_{k|k-1}\right]\left[\chi_{k|k-1}^i - \hat{X}_{k|k-1}\right]^T \tag{19}$$

$$P_{zz} = \sum_{i=0}^{2n} w_i^c \left[\zeta_{k|k-1}^i - \hat{Z}_{k|k-1}\right]\left[\zeta_{k|k-1}^i - \hat{Z}_{k|k-1}\right]^T + R_k \tag{20}$$

$$\begin{cases} K_k = P_{xz}P_{zz}^{-1} \\ \hat{X}_k = \hat{X}_{k-1} + K_k\left[Z_k - \hat{Z}_{k|k-1}\right] \\ P_k = P_{k|k-1} - K_k P_{zz} K_k^T \end{cases} \tag{21}$$

where K_k is the gain matrix, $\hat{X}_k$ is the state vector at time k, and P_k is the covariance matrix at time k.

3. Proposed Algorithm Based on the Maximum-Correntropy UKF

In this paper, based on the UKF, we first obtained the predicted state information and the covariance matrix through traceless transformation. Then, the nonlinear regression method based on the maximum cross-correlation entropy criterion was used to reconstruct observation information, which enhanced the robustness of the unscented Kalman filter algorithm for heavy-tailed noise.

3.1. Principle of Maximum Correntropy

The maximum entropy principle, also known as the maximum information principle, is a criterion for selecting the statistical characteristics of random variables that are the most consistent with an objective situation. The probability distribution of a random quantity is difficult to measure, and generally, only its various mean values, such as mathematical expectation and variance, or the values known under certain conditions. The distribution of these values can be varied or even infinite. Generally, one distribution will have the largest entropy. We select the distribution with the maximum entropy as the distribution of the random variable is an effective method and criterion [19].

Given two random variables, $\mathbf{X}$ and $\mathbf{Y}$, the correlation entropy between them is defined as [20,21]

$$V(\mathbf{X},\mathbf{Y}) = E[\kappa(\mathbf{X},\mathbf{Y})] = \iint \kappa(\mathbf{X},\mathbf{Y})F(\mathbf{X},\mathbf{Y})dxdy \tag{22}$$

where $E[\cdot]$ represents the expectation, $\kappa(\mathbf{X},\mathbf{Y})$ represents the kernel function and $F(\mathbf{X},\mathbf{Y})$ represents the joint probability density function of the random variables $\mathbf{X}$ and $\mathbf{Y}$. The authors of this paper selected the Gaussian kernel function as their kernel function:

$$\kappa(x_i, y_i) = G_\sigma(e_i) = exp\left(\frac{e_i^2}{2\sigma^2}\right) \tag{23}$$

where $\{x_i, y_i\}_{i=1}^N$ represents N samples meeting the joint probability density, $F(\mathbf{X},\mathbf{Y})$. $e_i = x_i - y_i$. σ represents the bandwidth of the kernel function.

Because only a limited part of the data in the actual system is known and it is difficult to accurately obtain joint probability density, the average value of the sample is usually used to calculate the estimated value of the correlation entropy:

$$V(\mathbf{X},\mathbf{Y}) = \frac{1}{N}\sum_{i=1}^{N}\kappa(x_i, y_i) \tag{24}$$

Through substitution of Equation (23) into Equation (24) and expanding of the Gaussian kernel function with the Taylor expansion, we derived

$$V(\mathbf{X},\mathbf{Y}) = \sum_{n=0}^{\infty}\frac{(-1)^n}{2^n\sigma^{2n}n!}E\left[(\mathbf{X}-\mathbf{Y})^{2n}\right] \tag{25}$$

The above formula shows that correlation entropy is obtained from the weighted sum of all even-order moments of the random variable $(\mathbf{X}-\mathbf{Y})$. Therefore, Formula (25) contains high-order moment information. At the same time, the correlation entropy, $V(\mathbf{X},\mathbf{Y})$, receives the maximum value of 1 only when $\mathbf{X}=\mathbf{Y}$. With selection of the right σ, higher-order moments can be incorporated into the signal-processing algorithm to more accurately describe the higher-order characteristics of the non Gaussian distribution. This is the main advantage of correlation entropy in processing non Gaussian noise.

3.2. Proposed Algorithm

This section uses the UWB positioning system as the research object. Aiming at the problem of outlier interference in NLOS environments, a novel unscented Kalman filter algorithm was designed with the maximum cross-correlation entropy criterion and the introduction of the Gaussian kernel function. The specific derivation process is as follows:

Combined with the maximum-entropy criterion and the MMSE criterion, the recursive expression of the UKF can be obtained via solving the following cost function:

$$J_{KF} = \min_{\mathbf{X}_k}\left(\|\mathbf{X}_k - \hat{\mathbf{X}}_{k|(k-1)}\|^2_{\mathbf{P}^{-1}_{k|(k-1)}} + \|\mathbf{H}_k\mathbf{X}_k - \mathbf{Z}_k\|^2_{\mathbf{R}^{-1}_k}\right) \tag{26}$$

where $\hat{\mathbf{X}}_{k|(k-1)}$ represents the state vector, $\mathbf{X}_k$ is the a priori estimate of k and $\mathbf{P}_{k|(k-1)}$ represents the prior error covariance matrix;

$$e_k = \mathbf{R}_k^{-1/2}(\mathbf{H}_k\mathbf{X}_k - \mathbf{Z}_k) \tag{27}$$

The above formula can be expressed as

$$J_{KF} = \min_{\mathbf{X}_k}\left(\|\mathbf{X}_k - \hat{\mathbf{X}}_{k|(k-1)}\|^2_{\mathbf{P}^{-1}_{k|(k-1)}} + \sum_{i=1}^{n} e_{ki}^2\right) \tag{28}$$

where e_{ki} is the component of k is $i = (1, \cdots, m)$. The formula above shows that under the MMSE criterion, the cost function of the Equation (28) assigns equal weight values to the residuals of all observations. However, the optimal estimation result cannot be obtained in the case of a measurement anomaly. Through introduction of the Gaussian kernel function in the maximum cross-correlation entropy criterion into the measurement of the cost function, the cost function of Equation (28) can be modified as follows:

$$J_{MCKF} = \min_{\mathbf{X}_k}\left(\|\mathbf{X}_k - \hat{\mathbf{X}}_{k|(k-1)}\|^2_{\mathbf{P}^{-1}_{k|(k-1)}} + \sigma^2\sum_{i=1}^{n} G_\sigma(e_{ki})\right) \tag{29}$$

Among these variables, $\sigma > 0$ indicates the bandwidth of the kernel function. Formula (29) replaces the MMSE estimation in the measurement-error section of the UKF with the kernel function of the maximum cross-correlation entropy.

For Formula (29), in X_k, the derivation at k gives:

$$P_{k|(k-1)}^{-1}\left(X_k - \hat{X}_{k|(k-1)}\right) + \sum_{i=1}^{n} G_\sigma(e_{ki})e_{ki}\frac{\partial e_{ki}}{\partial x_k} = 0 \tag{30}$$

The formula above is organized into matrix form as follows:

$$P_{k|(k-1)}^{-1}\left(X_k - \hat{X}_{k|(k-1)}\right) + H_k^T R_k^{-T/2}\psi_k e_k = 0 \tag{31}$$

Among these variables, $\psi_k = diag[G_\sigma(e_{ki})]$ indicates the weight matrix and $diag[\bullet]$ represents a diagonal matrix. Through changing of e_k via substitution of the expression of k into Formula (31), we derived

$$P_{k|(k-1)}^{-1}\left(X_k - \hat{X}_{k|(k-1)}\right) + H_k^T R_k^{-T/2}\psi_k R_k^{-1/2}(H_k X_k - Z_k) = 0 \tag{32}$$

The above equation can be regarded as Equation (33) for the X_k derivative at k:

$$J_{MCKF} = \min_{X_k}\left(\|X_k - \hat{X}_{k|(k-1)}\|^2_{P_{k|(k-1)}^{-1}} + \|H_k X_k - Z_k\|^2_{\tilde{R}_k^{-1}} \right) \tag{33}$$

where

$$\tilde{R}_k^{-1} = R_k^{-T/2}\psi_k R_k^{-1/2} \tag{34}$$

Comparison of Formula (28) and Formula (33) shows that the filtering algorithm based on the maximum cross-correlation entropy passes the weight matrix on the basis of the unscented Kalman filtering algorithm, ψ_k. The covariance of measurement noise is modified. The MCUKF method is as follows:

Step 1: Initialize the state vector and the covariance matrix.
Step 2: State one-step forecast update.
Step 3: State the one-step prediction mean square error update.
Step 4: According to Formula (34), update the measurement-noise covariance matrix.
Step 5: According to Formula (21), update the UKF measurement.

4. Results and Analysis

To verify the UKF algorithm based on the maximum cross-correlation entropy criterion proposed in this paper, a simulation and experimental verification were designed.

4.1. Simulation

The UWB positioning simulation platform include four base stations and one label, which was designed according to the UWB positioning principle. The local position frame, in which the positioning coordinates of the four base stations were BS1 (1 m,1 m), BS2 (14 m, 2 m), BS3 (16 m, 19 m) and BS4 (2 m, 21 m), respectively, was established. In terms of setting the ranging-error parameters, the distance measureing error from the base station to the label was ±15 cm. In the simulation process, labels moved uniformly in a straight line within the area enclosed in the four base stations. The starting position of the label was (1 m, 10 m), and the whole process lasted for 500 s. Details are shown in Figure 2. In order to demonstrate the effectiveness of the MCUKF method, the least squares method (LSM) was introduced to compare with the CKF.

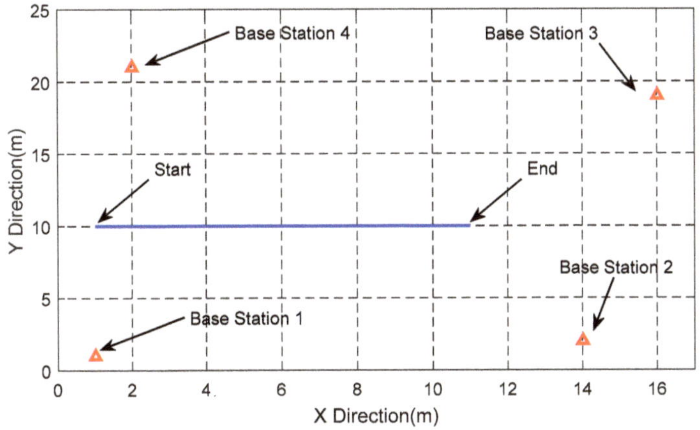

Figure 2. Simulation of carrier-movement track and base-station distribution.

Based on the simulation conditions above, the error curve of the positioning results was drawn, as shown in Figure 3. The LSM is marked with a black dashed line, the UKF method is marked with a blue line and the MCUKF method is marked with a red line. As can be seen from the figure, the MCUKF method proposed in this paper is superior to the LSM and the UKF method in positioning errors in both the X direction and the Y direction.

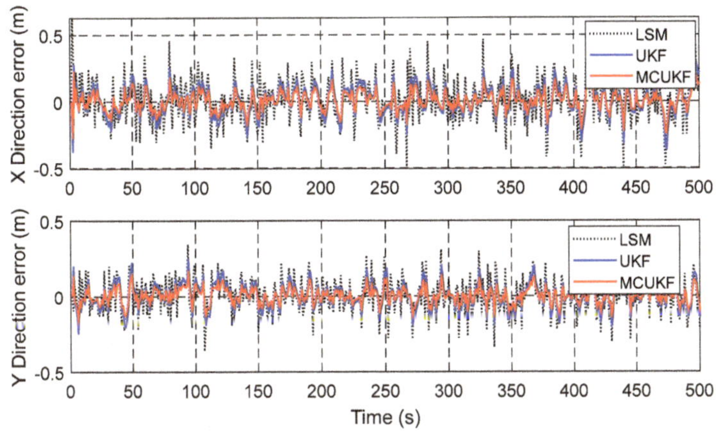

Figure 3. X− and Y−direction position-error simulation curve based on MCUKF method.

In order to comprehensively compare the positioning accuracy of the three methods, the horizontal position error curve, shown in Figure 4, was drawn. This figure shows that the horizontal positioning accuracy of the MCUKF method was better than that of the LSM and the UKF method.

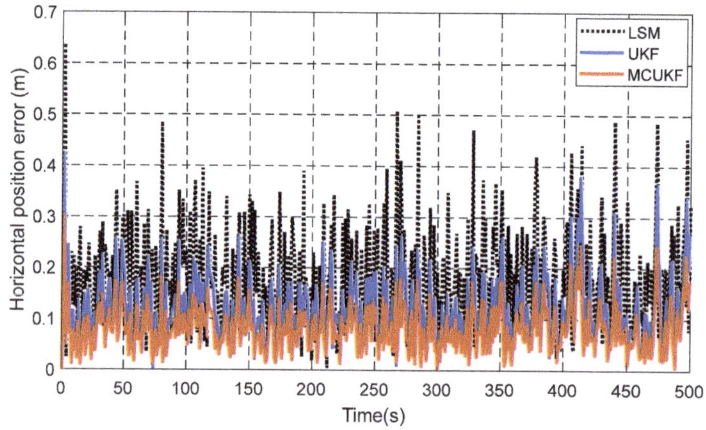

Figure 4. Horizontal position-error simulation curve based on MCUKF method.

The results in Figures 3 and 4 gave the statistical table of the root mean square error shown in Table 1. This table shows that the MCUKF method is superior to the LSM and the UKF method no matter the root mean square error of the X or Y direction or the root mean square error of the horizontal position. Compared with the LSM, the root mean square error results of the X direction, the Y direction and the horizontal position found with the MCUKF method decreased by 54.50%, 52.52% and 53.78%, respectively. Compared with the UKF method, the MCUKF method values were reduced by 28.89%, 31.76% and 31.34%, respectively. Therefore, the proposed MCUKF method has better positioning accuracy.

Table 1. Simulation of RMS of UWB positioning errors.

Method	X-Direction Error (m)	Y-Direction Error (m)	Horizontal Position Error (m)
LSM	0.1655	0.1190	0.2038
UKF	0.1059	0.0828	0.1372
MCUKF	0.0753	0.0565	0.0942

4.2. Test Verification

To verify the MCUKF method, a UWB positioning system based on the base station and the label was designed. The system included four base stations, one label, a computer and a trolley; the specific block diagram is shown in Figure 5. The computer was responsible for recording information about the distance from the label to the four base stations, as well as the positioning results.

The test system was built as shown in Figure 6, according to the framework above. The test area was an underground garage. One of the base stations was taken as the coordinate origin, and the four UWB base stations were each placed in a different corner. The label was placed on the trolley, and the trolley ran according to the specified route. The UWB update frequency was 2 Hz.

The whole test was divided into two parts: a static test and a dynamic test. It mainly verified the performance of the MCUKF algorithm under external interference. Figure 7 shows the distribution of the positioning results of the three methods when the label was stationary. The figure shows that due to the influence of external interference, the positioning results of the least squares method and the UKF method fluctuated. The MCUKF method reduced the external interference and had high positioning accuracy.

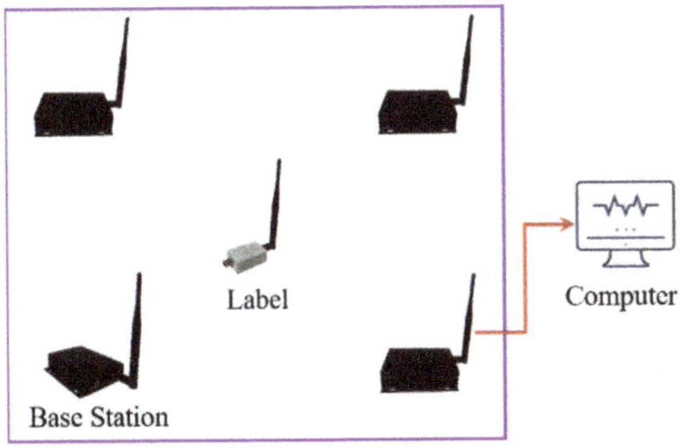

Figure 5. Block diagram of UWB positioning system.

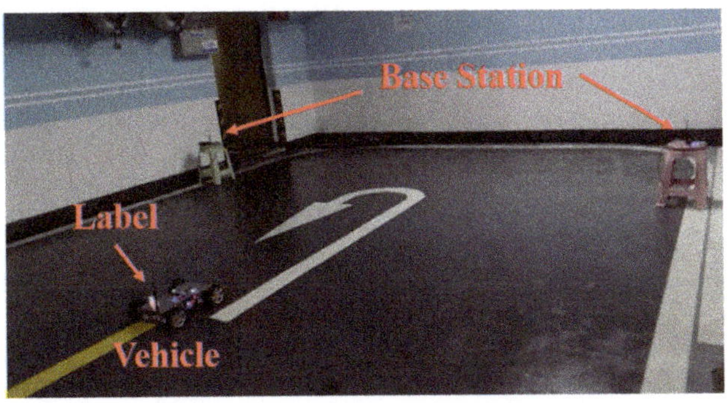

Figure 6. Field-test diagram of UWB positioning system.

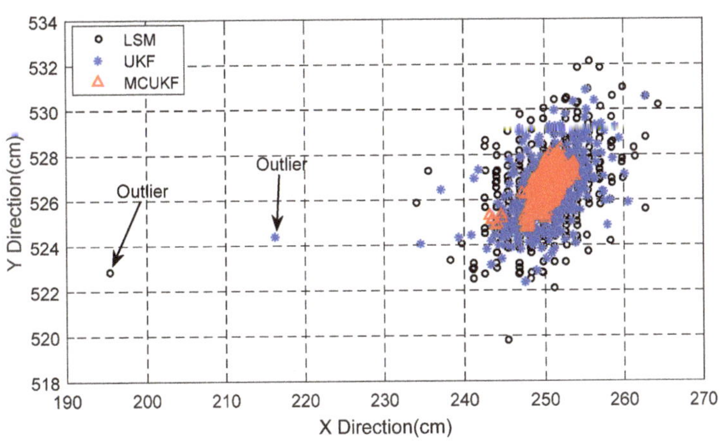

Figure 7. Comparison of positioning results of different methods under static conditions.

To facilitate analysis of the above results, the distribution curve of the standard deviation of the positioning results was drawn, as shown in Figure 8. The standard-deviation curves of the positioning results in the X and Y directions in the figure show that the standard deviation of the positioning results in the lowest multiplication method was the largest. The MCUKF method presented in this paper had the minimum standard deviation distribution.

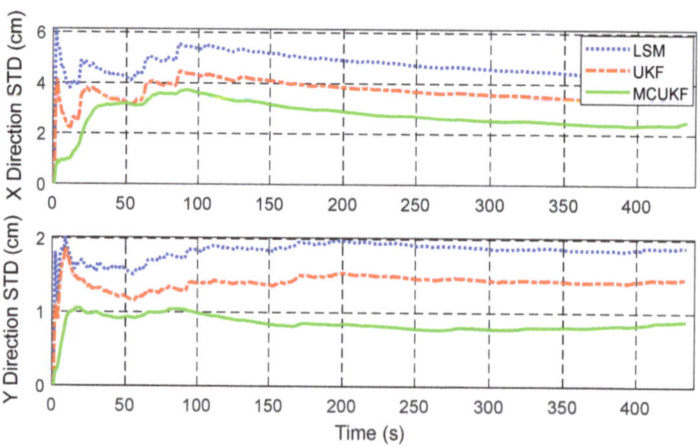

Figure 8. Distribution curve of standard deviation of positioning results.

Statistical analysis of the results in Figure 8 obtained the average-standard-deviation statistical results shown in Table 2. The average standard deviations of the MCUKF method proposed in this paper, in the X direction and the Y direction, were 1.6075 m and 0.7472 m, respectively. As seen in this table, the MCUKF method had the smallest mean standard deviation compared with the LSM and the UKF method. This table also shows that the proposed UWB localization method based on the MCUKF method is robust.

Table 2. Statistical results of mean standard deviation.

Method	X-Direction Error (m)	Y-Direction Error (m)
LSM	5.1034	1.8855
UKF	3.9037	1.4531
MCUKF	1.6075	0.7472

In order to illustrate the universal applicability of this method, the static-positioning test was redesigned. Figure 9 shows the UWB positioning results under the condition of double interference. The figure shows that when there was external interference, the positioning results of the least squares method and the UKF method both had outlier fluctuations. The MCUKF method proposed in this paper still maintained high robustness and positioning accuracy.

Figure 10 is the standard-deviation distribution curve under the condition of interference drawn on the basis of Figure 9. This figure shows that the standard-deviation curve also fluctuated greatly due to the influence of large outliers. The standard-deviation curve of the MCUKF method proposed in this paper remained stable under the influence of large outliers.

Similarly, the average standard-deviation statistical results of the different methods shown in Table 3 are listed according to the results in the figure above. The average standard deviations of the MCUKF method were 1.661 m in the X direction and 0.752 m in the Y direction, respectively. Meanwhile, compared with those of the LSM, the mean standard deviations of the method were reduced by 74.85% and 67.03%, respectively. Compared withUKF method, the average standard deviations decreased by 63.14% and 54.51%, respectively. The MCUKF method had high robustness.

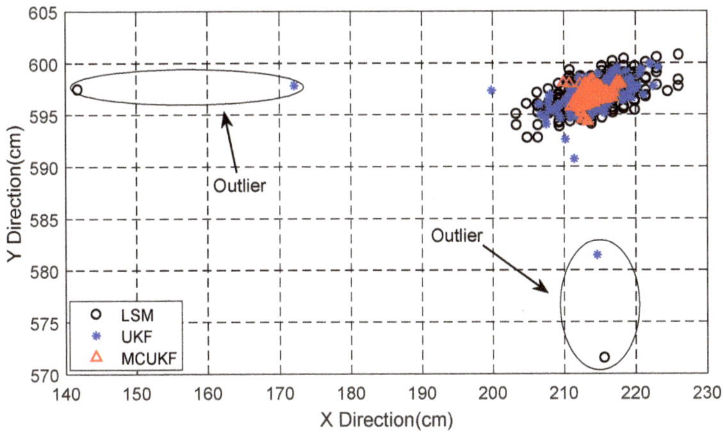

Figure 9. Test chart of UWB positioning results under static conditions.

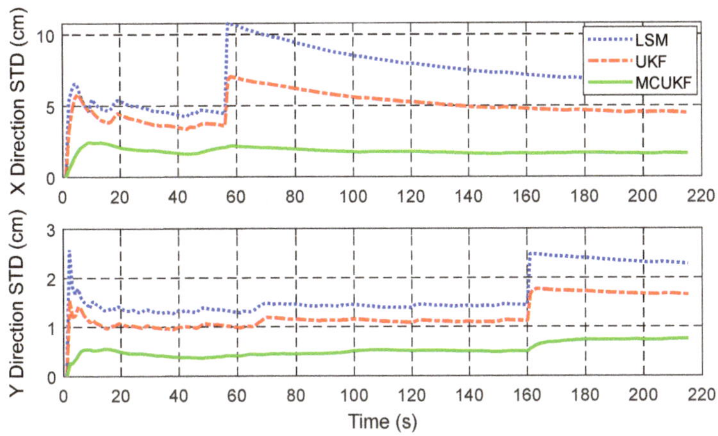

Figure 10. Distribution curve of standard deviation under interference.

Table 3. Statistical results of mean standard deviation for different methods.

Method	X-Direction Error (m)	Y-Direction Error (m)
LSM	6.603	2.281
UKF	4.506	1.653
MCUKF	1.661	0.752

In order to further illustrate the effectiveness of the proposed method, a dynamic UWB positioning test was designed. The moving area of the trolley was 14 m $\times$ 5 m, including acceleration, deceleration, straight lines and turning.

Figure 11 shows the positioning-result error curves of the three methods under dynamic conditions. The circle symbol represents the least squares method. The blue line represents the UKF method. The red line represents the MCUKF method proposed in this paper. This figure shows that the positioning result of the least squares method had large fluctuations due to external interference. Meanwhile, compared with the LSM, the positioning result of the UKF method was relative, but still displayed fluctuation. The MCUKF method proposed in this paper had the best robustness and high positioning accuracy.

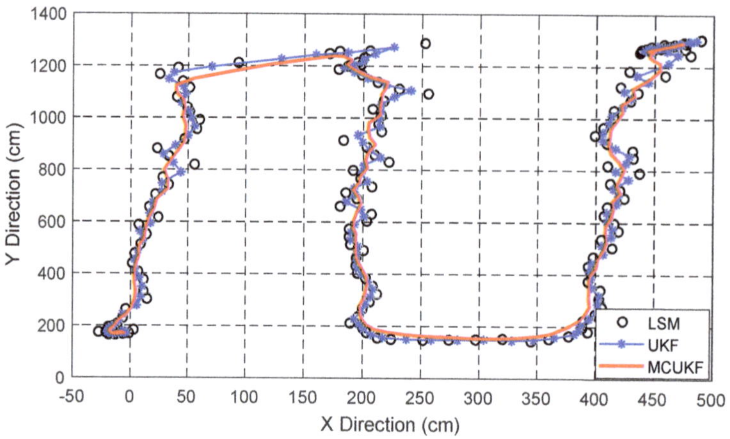

Figure 11. Curves of positioning results of various methods under dynamic conditions.

5. Conclusions

Aiming at the problem that UWB location is easily interfered with by outliers, this paper proposes a UKF algorithm based on maximum cross-correlation entropy. On the one hand, a UWB location model based on nonlinear filtering was derived. On the other hand, the predicted state estimate and the covariance matrix were obtained through traceless transformation, and the observation information was reconstructed using the nonlinear regression method based on the maximum cross-correlation entropy criterion to further improve the anti-interference ability of the filter. The simulation and vehicle test results show that the MCUKF algorithm proposed in this paper can maintain high robustness under the condition of wild value interference. At the same time, the UWB positioning accuracy of the algorithm proposed in this paper is better than that of the least squares method and the UKF method.

Author Contributions: Writing—original draft, M.Z.; Writing—review & editing, T.Z. and D.W. All authors have read and agreed to the published version of the manuscript.

Funding: This work was supported in part by the Key R&D program of Jiangsu Province under Grant BE2021679, the Key R&D and Transformation program of Qinghai Province under Grant 2022-QY-208 and the National Disabled Persons' Federation project under Grant 2021CDPFAT-26.

Institutional Review Board Statement: Not applicable.

Informed Consent Statement: Not applicable.

Data Availability Statement: Not applicable.

Conflicts of Interest: The authors declare no conflict of interest.

References

1. Barbieri, L.; Brambilla, M.; Trabattoni, A.; Mervic, S.; Nicoli, M. UWB localization in a smart factory: Augmentation methods and experimental assessment. *IEEE Trans. Instrum. Meas.* **2021**, *70*, 2508218. [CrossRef]
2. Jiang, C.; Shen, J.; Chen, S.; Chen, Y.; Liu, D.; Bo, Y. UWB NLOS/LOS classification using deep learning method. *IEEE Commun. Lett.* **2020**, *24*, 2226–2230. [CrossRef]
3. Jiang, C.; Chen, S.; Chen, Y.; Liu, D.; Bo, Y. An UWB channel impulse response de-noising method for NLOS/LOS classification boosting. *IEEE Commun. Lett.* **2020**, *24*, 2513–2517. [CrossRef]
4. Xu, A.; Liu, T.; Sui, X.; Wang, C.Q. Indoor positioning and attitude determination method of UWB/INS tight combination. *J. Navig. Position.* **2017**, *5*, 14–19.
5. Liu, T.; Xu, A.; Sui, X. UWB indoor location based on adaptive robust Kalman filter. *J. Sens. Technol.* **2018**, *31*, 567–572.
6. Li, R.; Wang, N.; Liu, J. Estimation and compensation method of UWB ranging error for relative navigation. *J. Instrum.* **2019**, *40*, 28–35.
7. Wang, R.; Zheng, S.; Chen, H. An indoor collaborative location method based on Taylor and Kalman. *J. Sens. Technol.* **2014**, *27*, 1557–1561.

8. Cai, G.; Xu, A.; Hong, Z. Robust volume Kalman filtering and its application in UWB indoor positioning. *Surv. Mapp. Sci.* **2018**, *43*, 123–129.
9. Huang, Y.; Zhang, Y.; Li, N.; Wu, Z.; Chambers, J.A. A novel robust Student's t-based Kalman filter. *IEEE Trans. Aerosp. Electron. Syst.* **2017**, *53*, 1545–1554. [CrossRef]
10. Liu, F.; Li, X.; Yuan, S.; Lan, W. Slip-aware motion estimation for off-road mobile robots via multi-innovation unscented Kalman filter. *IEEE Access* **2020**, *8*, 43482–43496. [CrossRef]
11. Alspach, D.; Sorenson, H. Nonlinear Bayesian estimation using Gaussian sum approximations. *IEEE Trans. Autom. Control* **1972**, *17*, 439–448. [CrossRef]
12. Xu, X.; Zhang, Y.; Ren, M.; Zhang, J.; Yan, G. Generalized correntropy filter-based fault diagnosis and tolerant control for non-Gaussian stochastic systems subject to sensor faults. *IEEE Access* **2018**, *6*, 12598–12607. [CrossRef]
13. Shi, W.; Li, Y.; Wang, Y. Noise-free maximum correntropy criterion algorithm in non-Gaussian environment. *IEEE Trans. Circuits Syst. II Express Briefs* **2019**, *67*, 2224–2228. [CrossRef]
14. Cinar, G.T.; Principe, J.C. Hidden State Estimation Using the Correntropy Filter with Fixed Point Update and Adaptive Kernel Size. In Proceedings of the 2012 International Joint Conference on Neural Networks (IJCNN), Brisbane, QD, Australia, 10–15 June 2012; pp. 1–6.
15. Huang, Y.; Xue, C.; Zhu, F.; Wang, W.; Zhang, Y.; Chambers, J.A. Adaptive recursive decentralized cooperative localization for multirobot systems with time-varying measurement accuracy. *IEEE Trans. Instrum. Meas.* **2021**, *70*, 8501525. [CrossRef]
16. Liu, X.; Qu, H.; Zhao, J.; Chen, B. Extended Kalman Filter under Maximum Correntropy Criterion. In Proceedings of the 2016 International Joint Conference on Neural Networks (IJCNN), Vancouver, BC, Canada, 24–29 July 2016; pp. 1733–1737.
17. Cao, Y.; Yang, C.; Li, R.; Knoll, A.; Beltrame, G. Accurate position tracking with a single UWB anchor. In Proceedings of the 2020 IEEE international conference on robotics and automation (ICRA), Paris, France, 31 May–31 August 2020; pp. 2344–2350.
18. Huang, Y.; Zhang, Y.; Zhao, Y.; Shi, P.; Chambers, J.A. A novel outlier-robust Kalman filtering framework based on statistical similarity measure. *IEEE Trans. Autom. Control* **2020**, *66*, 2677–2692. [CrossRef]
19. Landolsi, M.A.; Muqaibel, A.H.; Almutairi, A.F. UKF-based channel estimation and LOS/NLOS classification in UWB wireless networks. *J. Eng. Res.* **2016**, *4*, 15. [CrossRef]
20. Chen, B.; Wang, X.; Li, Y.; Principe, J.C. Maximum correntropy criterion with variable center. *IEEE Signal Process. Lett.* **2019**, *26*, 1212–1216. [CrossRef]
21. Huang, Y.; Zhang, Y.; Chambers, J.A. A novel Kullback–Leibler divergence minimization-based adaptive student's t-filter. *IEEE Trans. Signal Process.* **2019**, *67*, 5417–5432. [CrossRef]

applied sciences

MDPI

Article

The Full-Field Path Tracking of Agricultural Machinery Based on PSO-Enhanced Fuzzy Stanley Model

Yu Sun [1], Bingbo Cui [1,2,*], Feng Ji [1], Xinhua Wei [1,2] and Yongyun Zhu [1,2]

1 School of Agriculture Engineering, Jiangsu University, Zhenjiang 212013, China; 2222016051@stmail.ujs.edu.cn (Y.S.); 2222016028@stmail.ujs.edu.cn (F.J.); 1000003563@ujs.edu.cn (X.W.); yongyunzhu@ujs.edu.cn (Y.Z.)

2 Key Laboratory of Modern Agricultural and Technology, Ministry of Education and Jiangsu Province, Jiangsu University, Zhenjiang 212013, China

* Correspondence: cuibingbo@ujs.edu.cn

Abstract: The unmanned operation of agriculture machinery in the full field of farmland is an important part of unmanned farm and smart agriculture. Although the autonomous navigation for agriculture robot has been widely studied in literature, research on the full-field path tracking problem of agriculture machinery is rare. In this paper, in order to enhance the adaptivity of path tracking algorithm, an improved fuzzy Stanley model (SM) is proposed based on particle swarm optimization (PSO), where the control gain is modified adaptively according to the tracking error, velocity and steering actuator saturation. The PSO-enhanced fuzzy SM (PSO-FSM) is verified by experiments on numerical simulation and self-driving of mobile vehicle. Simulation results indicate that the PSO-FSM achieves a better result than SM and FSM, where PSO-FSM changes the control gain adaptively under different velocities and actuator saturation conditions, and the maximum lateral errors of SM and PSO-FSM for mobile vehicle autonomous turning are 0. 32 m and 0.03 m, respectively. When the location of the mobile vehicle deviates from the expected path at 4 m in a lateral direction, the distance of the guided trajectory for the mobile vehicle to reach the expected path is no more than 5 m. A preliminary experiment is also carried out for a wheeled combine harvester working on slippery soil, and the result indicates that the maximum lateral tracking error of PSO-FSM is 0.63 m, which is acceptable for the path tracking of a combine harvester with a large operation width.

Keywords: agriculture machinery; autonomous navigation; adaptive path tracking; Stanley model

1. Introduction

Intelligent agricultural machinery is a crucial support for unmanned farm and smart agriculture, where automatic navigation is the core technology and has been widely studied [1,2]. During the special period of the outbreak of the new crown epidemic, the automatic navigation of agricultural machinery provides an important guarantee for agricultural production and food security. The path tracking of agricultural machinery controls the agricultural machinery driving along a predetermined trajectory autonomously, which is an important part of the unmanned autonomous operation of agricultural machinery in the farmland [3,4].

The widely employed path tracking control methods for agricultural machinery include PID control, pure pursuit control, fuzzy control, neural network-based control and optimal control [5–9], to name a few. Hu et al. [10] proposed a cascaded navigation control method for straight line path tracking, which separates the navigation control into path tracking control and steering control, and the standard deviation of the lateral error is 0.04 m. Zhang et al. [11] developed a path tracking controller by integrating an adaptive neural network estimator and a saturated auxiliary system, and compared it with the traditional methods. The controller reduces the tracking error by more than 28%. A model predictive control-based path following the controller was developed for high-speed vehicle path tracking in [12], where the fluctuation of longitudinal velocity is compensated in the predicted horizon to reduce the state prediction error. The experimental results indicate that the controller not only reduces the lateral and heading deviation effectively but also maintains the control stability under large vehicle maneuver. Li et al. [13] proposed an adaptive sliding mode control method based on non-time reference and RBF neural network to handle uncertain disturbances. The experiments demonstrate that the proposed method not only improves the path tracking performance but also eliminates the chattering phenomenon when the wheeled mobile robot suffers from uncertain disturbances. In order to overcome the disadvantages of pure pursuit in high-speed path tracking, a model predictive active yaw control implementation of pure pursuit path tracking was proposed in [14], which improves the tracking performance at high speeds by accommodating the vehicle's steady-state lateral dynamics. Most of the abovementioned adaptive path tracking algorithms improve the tracking precision or robustness with respect to disturbances at the cost of computational complexity, which hinders their application from real-time path tracking application.

Stanley model (SM) is a nonlinear feedback function based on lateral tracking error, which generates the steer angle command by employing the relative geometric relationship between vehicle's pose and predefined trajectory. SM does not depend on the look-ahead distance as pure pursuit, and its lateral tracking error converges to zero exponentially [15]. The unmanned vehicle based on SM is the winner of DARPA's second unmanned vehicle challenge, whereas its application on the automatic navigation of agricultural machinery is rarely reported. The adaptive parameter tuning is an active research field for a geometric-based path tracking controller, and there is similar problem for SM especially when vehicle maneuver or road conditions vary from time to time. The look-ahead distance of pure pursuit is adjusted in [16] by using improved particle swarm optimization (PSO), which indicates when the speed of agricultural machinery is 0.7 m/s and the driving distance exceeds 5 m, the maximum lateral error is 0.02 m. Wang et al. [17] utilized the multi-population genetic algorithm to optimize the parameters of SM controller, and results show that the path tracking performance is improved by 41.72% and 48.61% for two typical tractor turning methods (U and Ω routes), compared with SM without optimization. Amer et al. [18] proposed a PSO-optimized fuzzy supervisory system to handle the various trajectories and speed for an armored vehicle, where the parameters of the modified SM were changed adaptively based on fuzzy inference and an optimal knowledge database.

Inspired by the work of [18], this paper proposes an improved path tracking control algorithm based on the fuzzy SM (FSM) and PSO to cater for variations in the agriculture robot in terms of speed and road conditions. In this contribution, the primary controller parameters of SM are adjusted by fuzzy inference at first, and then the control gain under different velocities and actuator saturation conditions are further modified by PSO. The fuzzy algorithm is used to adjust the control gain with respect to tracking errors, which not only improves the tracking precision for the automatic turning of agricultural machinery, but also reduces the distance of guided trajectory when the initial lateral error is large. The PSO is employed to further optimize the decision-making of the front wheel steer angle and improve the adaptivity of agricultural machinery towards different vehicle speeds and actuator saturation situations. The main contributions of this paper are as follows: (1) PSO-enhanced control gain adjustment is developed to cater for varying vehicle velocities and actuator saturation conditions; (2) a cascaded path tracking algorithm based on FSM and PSO is proposed for the full-field autonomous navigation of agriculture robot; (3) the autonomous navigation experiments based on numerical simulation, a mobile vehicle and a combine harvester are employed to verify the superiority of PSO-enhanced FSM (PSO-FSM).

The structure of this paper is arranged as follows. The fuzzy Stanley model is presented in Section 2. PSO-enhanced control gain adjustment is developed in Section 3. In Section 4, the PSO-FSM is verified by using numerical simulation, path tracking of a mobile robot and an unmanned combine harvester in a slippery soil condition. Finally, Section 5 concludes the conclusions of this work.

2. Fuzzy-Based Stanley Model

In the path tracking of agricultural machinery, the bicycle model is often used for kinematic analysis, where it is assumed that the agricultural machinery is running on a smooth road, and only longitudinal pressure is generated between the tires and the ground.

As shown in Figure 1, δ is the front wheel rotation angle, θ is the heading angle of the vehicle and L is the distance between the front and rear axles of the vehicle. Kinematic analysis of this model can be summarized as

$$
\begin{aligned}
x'(t) &= v \sin\theta \\
y'(t) &= v \cos\theta \\
\theta'(t) &= \frac{v \tan\delta}{L}
\end{aligned}
\quad (1)
$$

where v is the vehicle longitudinal speed; $x'(t)$ is the vehicle speed in the x-axis direction; $y'(t)$ is the vehicle speed in the y-axis direction; and $\theta'(t)$ is the vehicle angular velocity. The simplified bicycle kinematic model will be employed in our simulation.

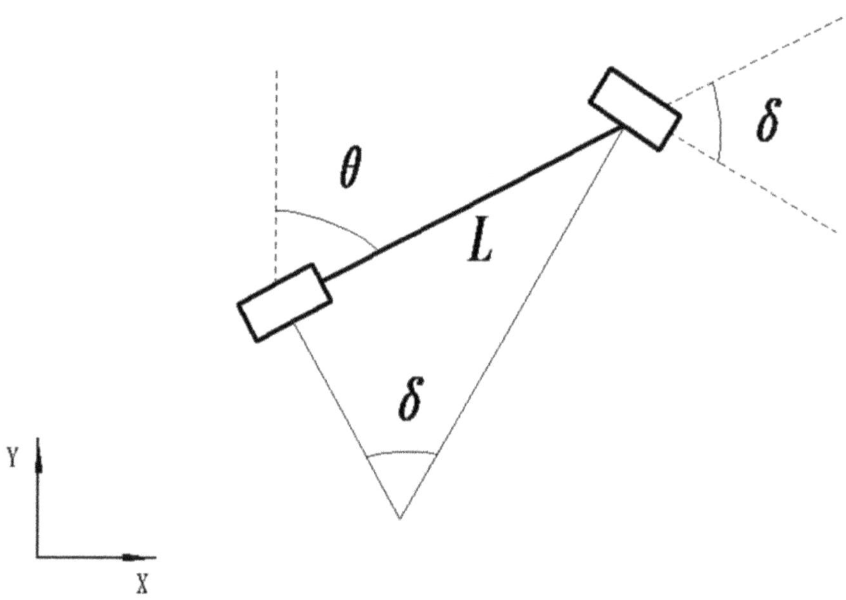

Figure 1. Vehicle kinematics model.

SM needs parameter adaptation similar to other geometric-based path tracking. Typical parameter self-adaptation methods include the intelligent search algorithm [19], neural networks algorithm [20] and fuzzy algorithm [21]. Intelligent search algorithms mainly include the genetic algorithm [22], PSO [23] and differential evolution algorithm [24]. The neural networks algorithm can be classified into BP [25], RBF [26], PNN [27] and GRNN [28] according to their excitation functions. Compared with other methods, parameter self-adaptation based on fuzzy algorithm has the advantages of low computational load and strong robustness, which is suitable for online implementation in an embedded controller [29]. In this section, the fuzzy algorithm is used to adjust the output gain of

SM, which provides an algorithm basis for global intelligent optimization under varying environmental disturbances.

2.1. Stanley Path Tracking Algorithm

The schematic diagram of SM is given in Figure 2, where the control input of the expected front wheel angle of the SM consists of two parts, i.e., corresponding to the lateral deviation and the heading deviation, respectively:

$$\delta(t) = \delta_e(t) + \delta_\theta(t) \tag{2}$$

where $\delta(t)$ is the expected angle; $\delta_e(t)$ is the expected angle due to lateral deviation; and $\delta_\theta(t)$ is the expected angle due to heading deviation.

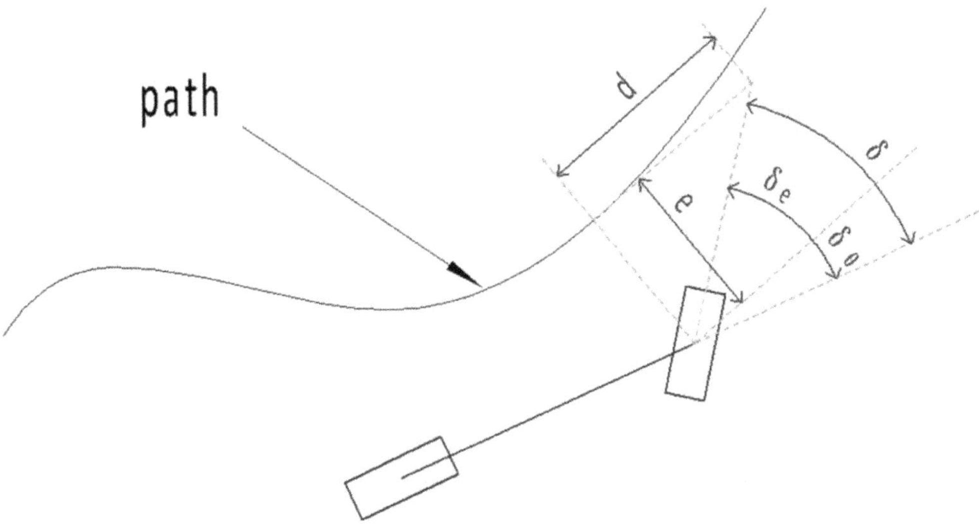

Figure 2. Schematic diagram of Stanley model.

By only considering the influence of lateral deviation, the larger the lateral deviation, the larger the expected front wheel steering angle. Suppose the expected path and the tangent line intersects at a point, that in front of the closest point with respect to front wheel by d, and according to the geometric relationship the following nonlinear scaling function can be obtained

$$\delta_e(t) = arctan\frac{e(t)}{d(t)} = arctan\frac{ke(t)}{v(t)} \tag{3}$$

where k is the gain coefficient; $e(t)$ is the lateral deviation; and $v(t)$ is the driving speed.

If only the influence of heading deviation is taken into consideration, the front wheel deflection angle is consistent with the tangent direction of the given path. The expected steering angle of the front wheel is equal to the angle between the vehicle heading and the tangent direction of the nearest path point, that is

$$\delta_\theta(t) = \theta_e(t) \tag{4}$$

where $\theta_e(t)$ is the heading deviation.

By taking both the two aspects into consideration, the expected rotation angle function of the front wheel is obtained as follows:

$$\delta(t) = \theta_e(t) + arctan\frac{ke(t)}{v(t)} \tag{5}$$

2.2. Parameter Self-Adaptation Based on Fuzzy Algorithm

It is notable in Equation (4) that the influence of the lateral deviation on the wheel steering angle is weighted by the gain coefficient of SM. The larger gain coefficient can reduce lateral deviation efficiently and enable the agricultural machinery to converge to the expected path quickly, whereas a large gain coefficient will lead to tracking error fluctuation. The small gain coefficient can make the agricultural machinery run smoothly and thus reduce the steady-state tracking error when the algorithm converges, but the controller converges slowly, especially when a large initial lateral error is given. Therefore, the time-varying gain coefficient should be selected according to different lateral deviations, especially for the full-field path tracking of agriculture machinery, which has stringent requirements on the distance of guided trajectory. What is more, the effect of the heading deviation and lateral deviation work on wheel steering angle is directional. When the lateral deviation and the heading deviation have the same direction of action on the wheel angle, the gain coefficient is decreased appropriately to make the controller stable. Conversely, when the lateral deviation and the heading deviation work on the wheel steering angle in opposite directions, the gain coefficient would be increased appropriately to speed up the algorithm's convergence speed.

The fuzzy reasoning rules are initially formulated by taking the actual conditions of agricultural machinery into consideration, where the membership function is constructed with the lateral deviation and heading deviation as input variables, and the steering angle of the front wheel as output whose maximum value is limited to 35 degrees. The maximum value and standard deviation of the lateral deviation are employed as observations, and a trial-and-error procedure is performed to find the optimal gain coefficient under different inputs.

The universe for the lateral deviation is [−3 m, 3 m], which includes right large, right medium, right small, zero, left small, left medium, left large, respectively, corresponding to NP, NM, NS, ZO, PS, PM and PB, as it is shown in Figure 3a. The universe of the heading deviation is [−30°, 30°], as it is shown in Figure 3b, which includes right large, right medium, right small, zero, left small, left medium and left large, respectively, corresponding to NP, NM, NS, ZO, PS, PM and PB. As it is shown in Figure 3c, the universe of gain coefficients is [0, 1.2], and the gain coefficients include zero, small, medium and large, respectively, corresponding to ZO, PS, PM and PB.

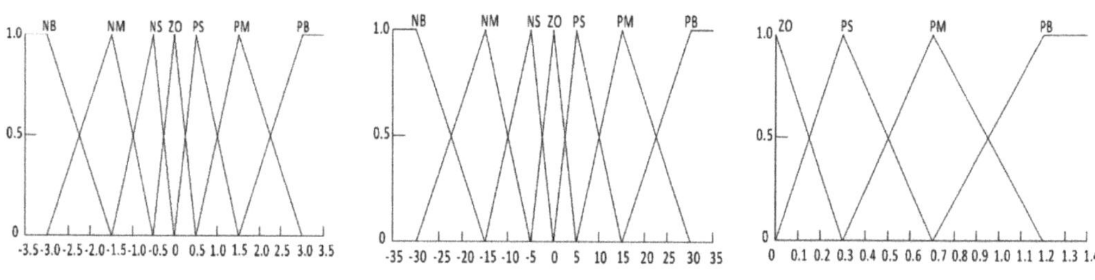

Figure 3. Membership function design for FSM. (**a**) Membership function of lateral deviation; (**b**) Membership function of heading deviation; (**c**) Membership function of gain coefficient.

The specific rules of the fuzzy inference table are designed as follows: the smaller the heading deviation and the larger the lateral deviation, then the larger the gain coefficient; if the heading deviation and the lateral deviation work on the wheel angle in the same direction, then the larger the heading deviation, the smaller the gain coefficient; if the heading deviation and lateral deviation work on the wheel angle in opposite directions, then the larger the heading deviation, the larger the gain coefficient; if the heading deviation is large, and the heading deviation and lateral deviation work on the front wheel angle in the same direction, then a smaller gain coefficient is selected; and if the heading deviation is large, and the heading deviation and lateral deviation work on the steering angle of the front wheel in opposite directions, then a larger gain coefficient is selected. Finally, after verification by experiment test, a total of 49 fuzzy rules are summarized in Table 1, where e stands for the lateral deviation and θ_e represents the heading deviation.

The typical defuzzification methods include maximum subordination principle, center of gravity method and weighted average method [30]. The center of gravity method is widely used in industry application because of its smooth output, so the center of gravity method is chosen in our work for defuzzification, which is formulated as

$$u_0 = \frac{\sum_{k=1}^{m} u_k \mu_k(u_k)}{\sum_{k=1}^{m} \mu_k(u_k)} \tag{6}$$

where μ_0 is the output coefficient; m is the number of levels; μ_k is the coefficients for each level; and $\mu_k(u_k)$ is the degree of membership.

Table 1. Rule base for fuzzy logic control.

θ_e	e						
	NB	**NM**	**NS**	**ZO**	**PS**	**PM**	**PB**
NB	PS	PS	PS	PM	PB	PB	PB
NM	PM	PS	PS	PS	PM	PB	PB
NS	PM	PM	PS	PS	PM	PM	PB
ZO	PM	PM	PS	PS	PS	PM	PM
PS	PB	PM	PM	PS	PS	PM	PM
PM	PB	PB	PM	PS	PS	PS	PM
PB	PB	PB	PB	PM	PS	PS	PS

3. PSO-Enhanced Fuzzy Stanley Model

3.1. Principle of PSO Optimization

PSO is a heuristic swarm intelligence algorithm, which stems from the research on the predation behavior of bird flocks [31]. In order to find the place where most food is located, the flocks adjust their search directions by using their own experience and communication among the flocks. Because of its fast convergence speed, easy online implementation and simple principle, it is widely used in the fields of fuzzy control system design and other high-dimensional optimization problems [32]. In the application of intelligent agricultural machinery for full-field unmanned operation, the various operation speeds, soil conditions and actuator saturation of steering control system will have a great effect on the accuracy and stability of path tracking. Therefore, in order to improve the adaptively of fuzzy-based SM, the PSO algorithm is employed to further optimize the expected steering angle.

The steering control system of agricultural machinery is a closed-loop PID control system including an electric steering wheel, a mechanical steering mechanism and angle sensor fixed on the front wheel. It is easy to understand that when the driving speed is high, the steering wheel control input should be changed slowly and a small steering angle should be given. In order to handle the time-varying speed, PSO is employed to optimize the output gain of FSM. Define a control gain coefficient α, and the control gain of FSM is multiplied by the coefficient

$$\delta_{PSO}(t) = \alpha \cdot \delta(t) \tag{7}$$

where $\delta_{PSO}(t)$ is the desired front wheel angle after utilizing PSO.

When the speed of the agricultural machinery is high, the value of the coefficient α should be reduced to improve the stability of the control. What is more, the steering control systems of the wheeled mobile vehicle and combine harvester involved in our work are actually a first-order system with time delay, and the varying parameters of the closed-loop PID will have a great effect on the navigation control decision. When the mobile robot is subject to its steady-state stage of path tracking, i.e., at the end of a straight line tracking, the actuator saturation of the steering control system will make it difficult to respond to the large, expected input angle in a timely manner. However, by multiplying the expected steering angle by a coefficient α, the response speed of the steering control system can be improved, and thus the accuracy of the path tracking during the path transition period can be ensured.

3.2. Implemention of PSO-Enhanced FSM

The fitness function of PSO is also named as the objective function. The selection of fitness function has a significant effect on the convergence speed and steady-state accuracy of the PSO-FSM. The integral time absolute error (ITAE) is frequently applied as the performance metric in the automatic control community, which calculates the integral of the product of the time and absolute error value for some predefined time interval [33]. In this paper, the inverse of the sum of ITAE of lateral deviation and heading deviation is used as the fitness function, and the calculation formula is as follows

$$f = \omega_1 \Delta t \sum_{j=1}^{N} t_j |e(j)| + \omega_2 \Delta t \sum_{j=1}^{N} t_j |\theta_e(j)| \tag{8}$$

$$F = \frac{1}{f} \tag{9}$$

where f is the sum of ITAE of lateral deviation and heading deviation, ω_1 and ω_2 are the weights which correspond to ITAE, Δt is the integration interval, and $e(j)$ and $\theta_e(j)$ correspond to the lateral deviation and heading deviation at time $j\Delta t$. N is the number of control periods involved in the calculation of the f value, and F is the fitness value employed for PSO-FSM.

The PSO simulates birds in a flock by designing a massless particle where only the speed and position are adjustable, and the flowchart of PSO is shown in Figure 4. Firstly, each particle searches for the optimal solution individually in the search space, then, its position is recorded as the local optimal position. Secondly, each particle shares its optimal position with other particles in the entire flock to find the temporary global optimal position of the current flock. Finally, each particle in the flock adjusts its attributes continuously according to the optimal position of the individual and the current flock until the global optimal solution is found. The specific procedures are as follows:

(1) Initialize the particle flock. Parameters such as the inertia coefficient ω and learning factor c_1, c_2 are determined according to the operation of agricultural machinery.
(2) Calculate the fitness of each particle in the flock and update the optimal position of each particle, then calculate the optimal position of the current flock. Adjust the speed and position of each particle accordingly. The calculation formulas are as follows:

$$v_i(k+1) = \omega v_i(k) + c_1 \xi (p_{ibest}(k) - x_i(k)) + c_2 \eta \left(p_{gbest}(k) - x_i(k) \right) \tag{10}$$

$$x_i(k+1) = x_i(k) + v_i(k+1) \tag{11}$$

where $v_i(k)$ and $x_i(k)$ are the velocity and position of particle i in the k-th iteration, p_{ibest} is the optimal position of particle i, p_{gbest} is the optimal location of the entire flock and ξ and η are random numbers between 0 and 1.
(3) The particles are continuously updated until the maximum number of iterations is reached or the required convergence accuracy is achieved. Then, the optimal position of the flock is the value of the optimization coefficient α.

Finally, the input and output block diagram of PSO-FSM can be summarized in Figure 5, where e represents the lateral deviation, θ_e represents the heading deviation and δ_α represents the expected front wheel angle after PSO.

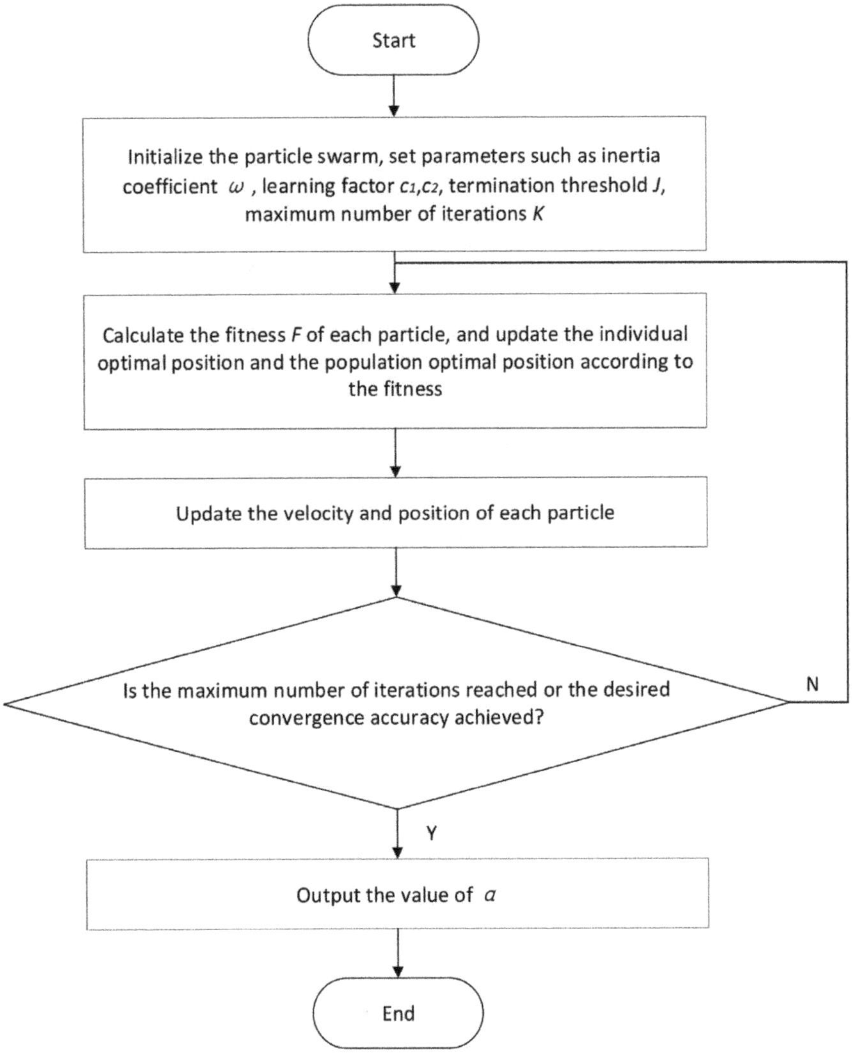

Figure 4. Flowchart of the PSO algorithm.

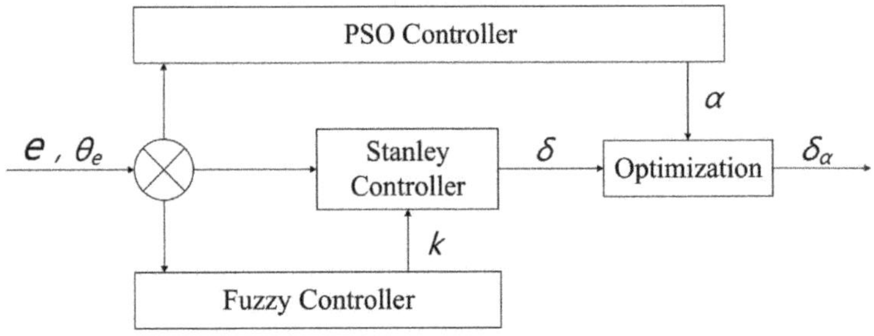

Figure 5. Input and output block diagram of PSO-FSM.

4. Results and Discussion

4.1. Numerical Simulation

In this section, the path tracking algorithms including SM, FSM and the proposed PSO-FSM are verified in autonomous turning under different scenarios using numerical simulation. A typical U turning trajectory is stimulated to verify the superiority of PSO-FSM. The flowchart of the numerical simulation is shown in Figure 6, where the vehicle kinematic model defined by (1) is employed to update the vehicle state, and the particle flock size is 20, the inertia coefficient $\omega = 0.5$, the learning factors $c_1 = 1$, $c_2 = 2$, the maximum number of iterations $K = 200$ and the termination threshold $J = 0.8$. These parameters not only reduce the occurrence of large errors but also prevents the overshoot of the system output.

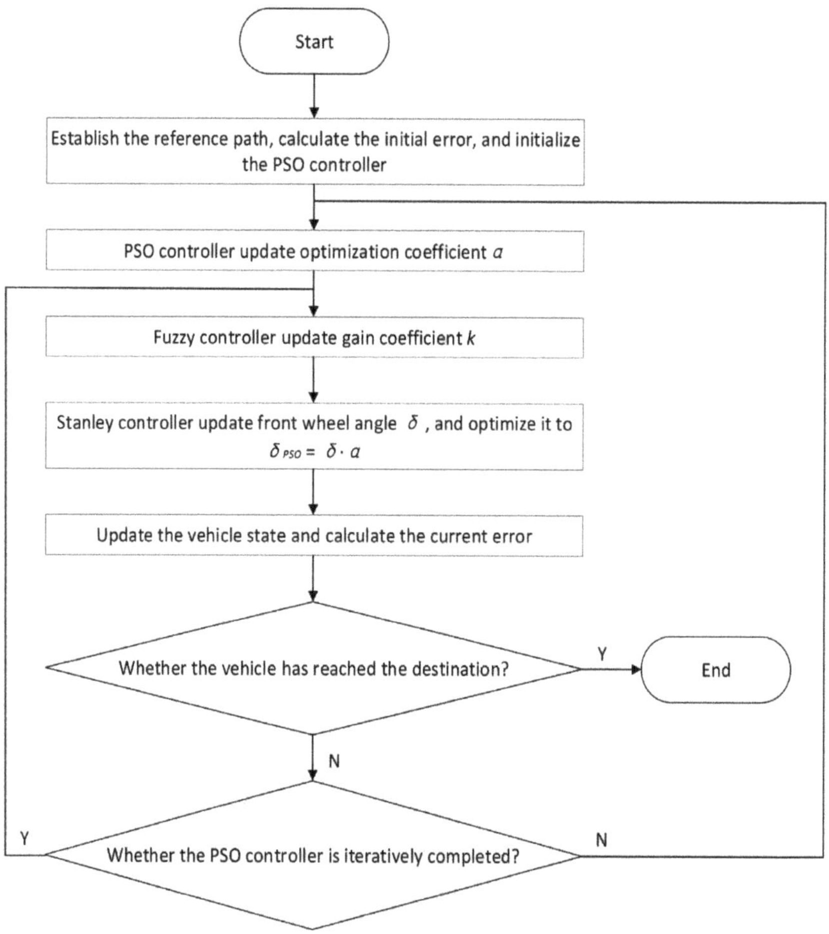

Figure 6. Flowchart of simulation experiment.

Figure 7a shows the test result when the vehicle runs at 1 m/s with small time delay of steering control system, which is an ideal work condition for agricultural machinery. It can be seen that both FSM and PSO-FSM achieve better performance than SM, especially when the curvature is variable, whereas PSO-FSM does not outperform FSM, obviously because the gain coefficient α does not have a direct modification on the expected front wheel angle. Figure 7b shows the results in the high-speed operation scenario, where the vehicle speed is increased to 3 m/s with other conditions unchanged. It is notable that the PSO-FSM achieves the best accuracy among the involved algorithms, and both SM and FSM have an obvious performance degradation. Once a low driving speed is given, e.g., 1 m/s, the actuator saturation situation and state delay of the control signal would have a great effect on the steering control system, which suits the actual operation scenario. By importing the control input limitation to the steering control system, i.e., scale the expected steering angle every control period by multiplying the random coefficient between 0 and 1, the steering actuator saturation can be stimulated. The corresponding result is shown in Figure 7c, which indicates when the vehicle starts turning with obvious curvature variation, PSO-FSM not only improves the response speed of steering control system, but also achieves better stability once the autonomous turning is finished. The detailed simulation results are listed in Table 2, where the mean absolute error (MAE) and root mean square error (RMSE) are taken as the performance metrics. As we can see, PSO-FSM achieves a much better result than FSM and SM, especially when the scenarios of high-speed or random actuator saturation appear.

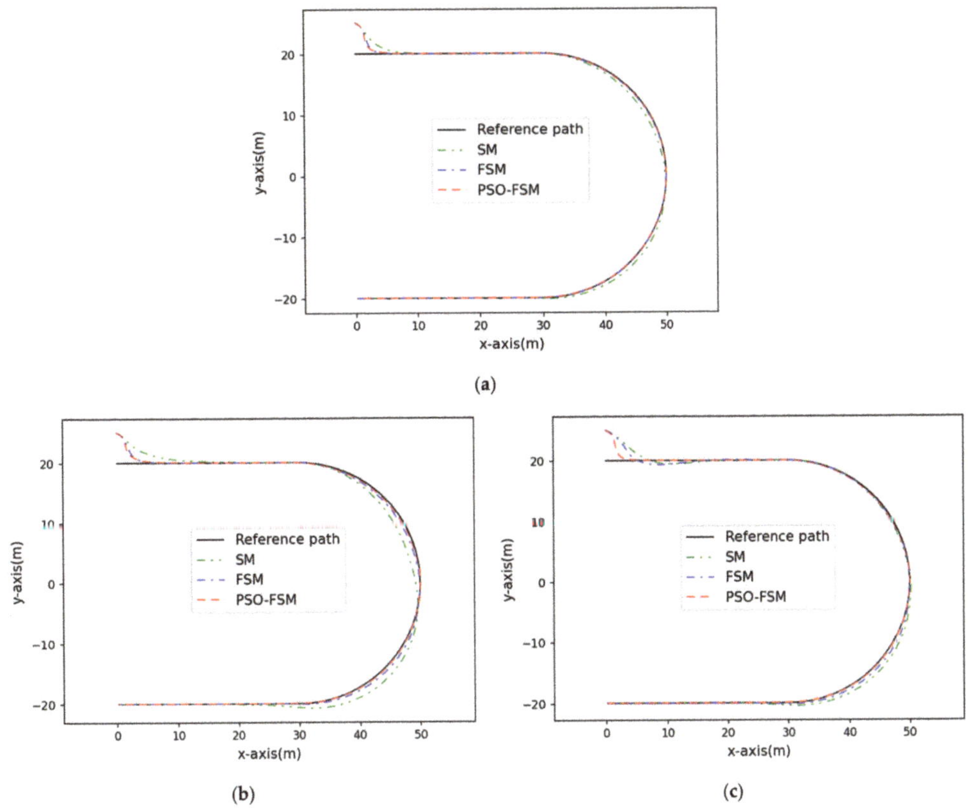

Figure 7. Autonomous turning under different scenarios. (**a**) Autonomous turning under ideal scenario; (**b**) Autonomous turning under high-speed scenario; (**c**) Autonomous turning under steering actuator saturation scenario.

4.2. Mobile Vehicle Autonomous Navigation Test

In order to verify the effectiveness of the proposed PSO-SM, a real-time autonomous navigation test is performed by employing a mobile vehicle, as shown in Figure 8. The dual-antenna RTK position system is applied to provide the position, velocity, heading and pitch information of the mobile vehicle. The base station receiver is OEM719 and the mobile station receiver is OEM718D (provided by NovAtel Inc., Calgary, AB, Canada), where the positioning accuracy is 1 cm + 1 ppm and heading accuracy is 0.08° with a 2 m antenna baseline. The update frequency of the dual-antenna RTK is 5 Hz. The motor driver RMDS405 supports 30 A current (provided by Shenzhen RoboModule Technology Co. Ltd., Shenzhen, China), and the encoder is Omron E6B2 with a resolution up to 3600 P/R.

Table 2. Path tracking result under different scenarios.

Velocity (m/s)	Steering Angle Scaling	Algorithm	MAE (cm)	RMSE (cm)
1	1	SM	1.1	36.1
1	1	FSM	0.4	13.6
1	1	PSO-FSM	0.3	12.4
3	1	SM	9.2	93.7
3	1	FSM	1.8	28.7
3	1	PSO-FSM	0.9	15.6
1	0~1	SM	20.9	48.1
1	0~1	FSM	5.4	16.2
1	0~1	PSO-FSM	0.3	12.5

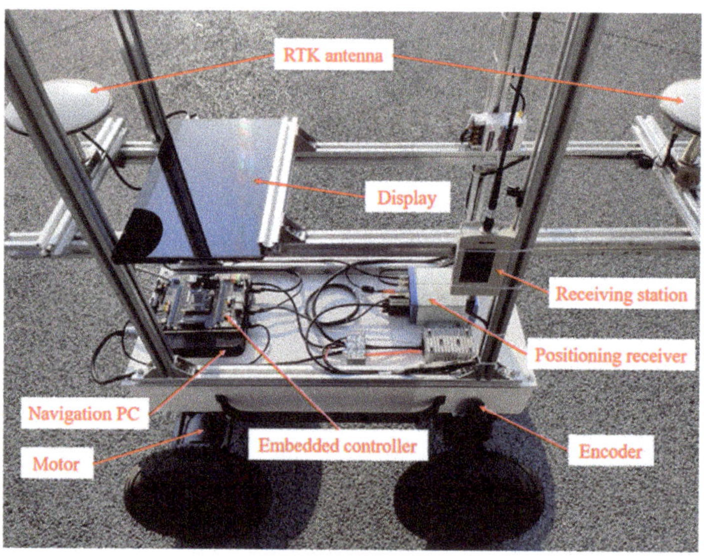

Figure 8. Setup for mobile vehicle path tracking test.

The navigation decision system NavLight (with Version 4.07) is built by the intelligent agricultural machinery team of Jiangsu University (Zhenjiang, China), which includes a full-field path plan and navigation signal processing and the cross-platform navigation software based on Qt works well by using multi-thread coding technology. First, predefined path information is generated utilizing the position information of the farmland. Second, the real-time heading and position information is sent to navigation PC and then lateral deviation and heading deviation are calculated and sent to the embedded controller. Finally, the controller calculates the desired steering angle based on the navigation deviation and sends the control information to the motor driver via the CAN bus at a frequency of 5 Hz. The motor driver converts the control command into voltage and receives feedback from the encoder to form a closed-loop control system, which can control the motor at 100 Hz.

In order to verify the adaptability of PSO-FSM to different path curvatures and speeds of agricultural machinery, the full-field working path for the combine harvester is designed.

As it is shown in Figure 9, a hybrid path is designed for the combine harvester of large operating width, where the outer circle is composed of autonomous turning path and the arrows indicate the driving direction. If half of the remaining field width is less than the turning radius of the combine harvester, the turning at the field head is carried out by manual driving, which is more efficient than autonomous turning with back-up control. The combine harvester performs harvesting operations in sequence. When the remaining field width is smaller than the size required for the harvester to carry out autonomous turning, the navigation PC sends a signal to the controller before the end of last straight line for autonomous turning, e.g., straight line 4 in Figure 9, and then automatically switches to the manual-assisted driving mode.

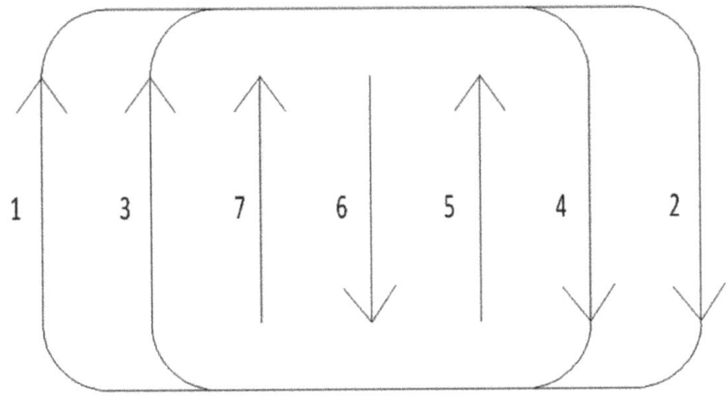

Figure 9. Full-field path planning schematic of the combine harvester.

The full-field path tracking of the combine harvester can be divided into the guiding stage, straight line and curve path tracking stage. The lateral tracking error of SM for the straight line is small, and the fixed output gain can achieve acceptable result, thus the guiding and curve path tracking stages are our main focuses in this work. The initial parameter setting of SM and PSO-FSM is the same as the numerical simulation to make a fair comparison. The average speeds of the mobile vehicle for the straight line and autonomous turning stages are 2.5 m/s and 0.8 m/s, respectively, i.e., the vehicle reduces its speed automatically before starting the curve path tracking. The path tracking trajectories of different algorithms for one U turning are shown in Figure 10, where both the guiding trajectory and curve path tracking result are included. It is notable in Figure 10a that PSO-FSM reduces the guiding distance of the mobile vehicle when the initial lateral deviation is set as 4 m, where the guiding distance is less than 5 m. In Figure 10b, because of the time-varying tracking error and steering actuator saturation, the performance of SM is degraded a lot compared with the straight line tracking. It is important to note that PSO-FSM outperforms SM which coincides with the numerical simulation, i.e., SM cannot achieve stability and precision path tracking by use fixed gain coefficient for time-varying curvature and disturbance. PSO and fuzzy inference improve SM significantly by employing a cascaded optimization structure to change the gain coefficient adaptively. The parameter variation in PSO-FSM is shown in Figure 11, where Figure 11a is the gain coefficient k of the path tracking test based on PSO-FSM. It is notable that k is updated adaptively according to the deviation change, and its value is reduced quickly when the path tracking of the curve segment is finished. Figure 11b is the expected front wheel steering angle calculated by the embedded controller. It is notable that when the curvature is changed, the expected steering angle changed accordingly, which presents different results with respect to the output gain coefficient variation.

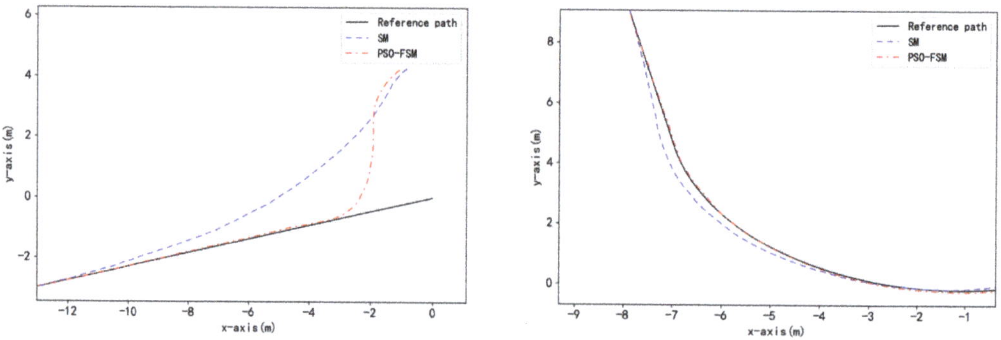

(**a**) (**b**)

Figure 10. Path tracking result under different scenarios. (**a**) Guiding trajectory with large initial lateral error; (**b**) Curve path tracking of autonomous turning.

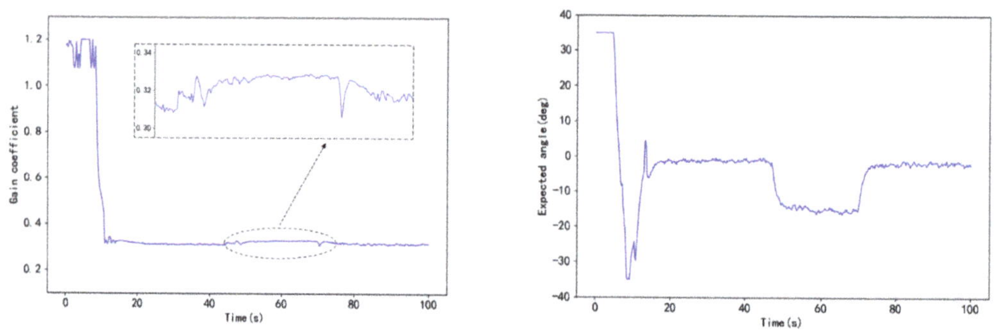

(**a**) (**b**)

Figure 11. Parameters of the path tracking based on PSO-FSM. (**a**) Gain coefficient of the path tracking; (**b**) Expected steering angle of the path tracking.

The tracking error of the different algorithms is shown in Figure 12, where the lateral errors of the straight line and curve segment are given. Notice that PSO-FSM reduces the guiding distance of the autonomous vehicle in Figure 12a, whereas once the path tracking is stable, i.e., after the vehicle runs after about 5 s, there is no obvious difference between SM and PSO-FSM. Both the algorithms achieve a maximum lateral tracking error less than 3 cm with a standard deviation less than 1 cm. However, the steady-state error of SM in the curve segment is improved significantly by PSO-FSM, where the maximum lateral error is reduced from 32 cm to 3 cm. In order to verify the full-field unmanned driving ability of the proposed algorithm, the full-field path tracking result of the mobile vehicle is shown in Figure 13, and the detail tracking errors are listed in Table 3. As we can see, PSO-FSM not only improves the guiding performance of the autonomous vehicle but also reduces the tracking error significantly. The maximum tracking error of PSO-FSM is less than 3 cm in the full-field path tracking test, which results from the fact that the uncertain disturbances of the mobile vehicle path tracking on a dry pavement is well handled by the cascaded path tracking algorithm.

When analyzing the experiment data, we notice that there is an obvious steady-state error in the straight line segment, which may result from the mechanical clearance of the steering control system and misalignment error between GNSS baseline and the forward direction of the mobile vehicle. In the application of large agricultural machinery, the error coming from the mechanical clearance of the steering control system is negligible and the misalignment error can be further compensated by field calibration [34].

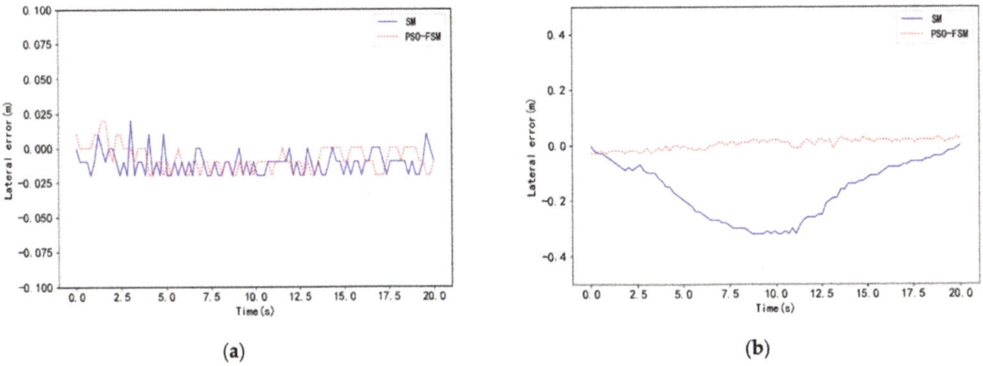

Figure 12. Path tracking result of different algorithms. (**a**) Lateral error of path tracking for straight line; (**b**) Lateral error of path tracking for curve path segment.

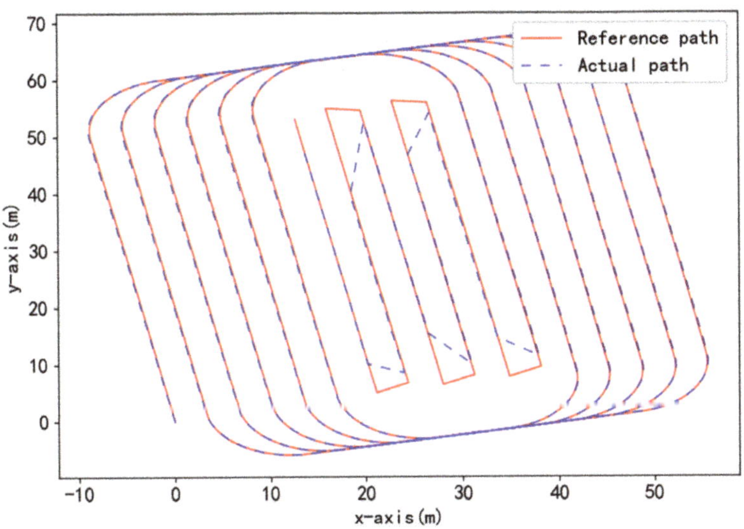

Figure 13. Full-field path tracking trajectory of PSO-FSM.

Table 3. Full-field path tracking error of the mobile vehicle.

Algorithm	Guiding Distance (m)	Maximum Error (cm)	MAE (cm)	RMSE (cm)
SM	12	32	7.4	11.4
PSO-FSM	5	3	1.6	2.7

4.3. Combine Harvester Field Test

In order to further verify the fitness of the PSO-FSM algorithm, a field test was performed based on a wheeled combine harvester C230 (provided by John Deere, Moline city, IL, USA) as shown in Figure 14. In the design of the autonomous navigation system of C230, the same dual-antenna RTK position system was employed as the wheeled mobile vehicle, and steering angle sensor, electric steering wheel were installed to set up the close-loop control of self-driving. Electric control ignition, electric control shift, electric control accelerator and electric control clutch were designed to realize unmanned combine harvester. The C230 harvester has an operating width of 5 m and a turning radius of about 8 m. Its navigation system configuration and operation during harvesting operation is consistent with that of the mobile vehicle experiment, and the path planning scheme shown in Figure 7 is employed for the full-field path tracking test.

Figure 14. Reconstruction of the electric control system for the combine harvester.

The whole control system includes an electric steering wheel, angle sensor, electric throttle and control unit related to grain unloading and threshing. The electric steering wheel is Shanghai Lianshi EMS2 steering drive unit, whose working current goes up to 10 A. The advantages of EMS2 include high torque, high precision, IP65 dustproof and waterproof, which is suitable for the harsh operating environment of a combine harvester. The angle sensor is DWQCAB-V-CH from Beijing Tianhaike Company, with a linearity of 0.02%FS and angular resolution of 0.022°. The angle sensor is installed directly above the steering vertical axis of the harvester and is directly connected to the center steering vertical axis. Because of the large operating width of C230, the average operating speeds for a straight line and automatic turn are about 0.8 m/s and 0.6 m/s, respectively, which is almost the same as manual driving operating. Limited by the large turning radius and reliability of manual–automatic gear switching, only the straight line harvesting operation and automatic turning at the outer circle path were performed in the field test.

The result of the unmanned operation of the combine harvester in the farmland is shown in Figure 15, where both the straight line and curved path tracking errors are smaller than 0.5 m if the path tracking algorithm is stable. However, the maximum lateral tracking error of the harvester is 0.63 m due to the harsh field soil environment, such as side slip resulting from field ridge. However, the tracking accuracy is acceptable for the C230 automatic navigation because of its large operation width. Compared with the mobile vehicle test result, we can conclude that the harsh soil condition and state delay of the steering control system have a complicated effect on the parameter tuning of SM, especially for bigger and heavier combine harvesters. Because heading information is only employed to correct the positioning error in this test, the lever-arm error coming from pitch and roll is not compensated, which can be seen from the field ridge-induced tracking error. Furthermore, the lever-arm of the GNSS antenna installed in C230 is obtained manually by measuring tape, and more precise lever-arm and high-frequency attitude information would improve the dual-antenna RTK position, which needs further verification and is also part of our future work.

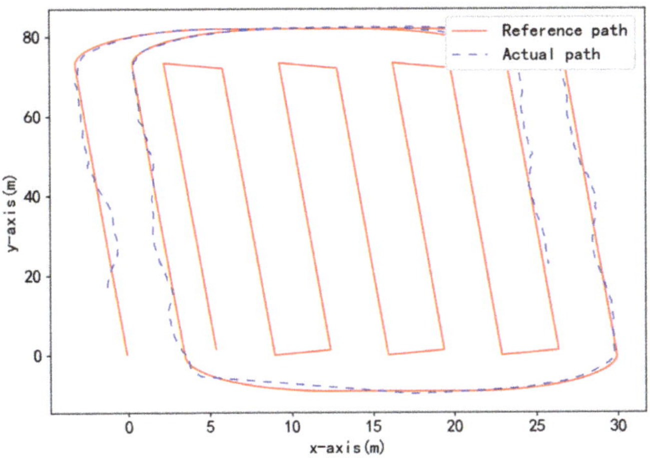

(a)

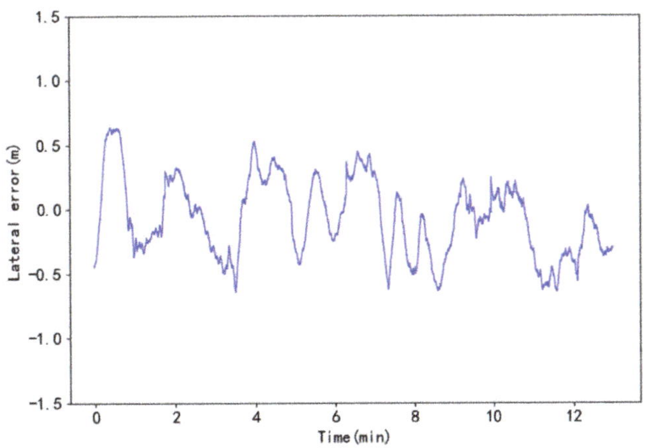

(b)

Figure 15. Path tracking result of the combine harvester. (**a**) Tracking trajectory of the combine harvester; (**b**) Lateral error of path tracking for the combine harvester.

5. Conclusions

In order to cater for the variation factors of the path tracking algorithm for autonomous agriculture machinery, the PSO-enhanced fuzzy Stanley model (PSO-FSM) is developed and verified. The internal factors of the path tracking control system are handled by using fuzzy algorithm design, and the time-varying external factors, such as actuator saturation and varying speed, are taken into consideration by PSO. Experiments based on numerical simulation and autonomous navigation vehicle are firstly performed, and the result demonstrates that PSO-FSM outperforms FSM and SM in terms of guiding distance under large initial error and path tracking precision under automatic turning. The field test of autonomous navigation vehicle indicates that the full-field tracking errors of PSO-FSM including straight line and curve path tracking are less than 3 cm. Then, a preliminary experiment on harsh farmland soil is performed, and the PSO-FSM achieves maximum tracking error at 0.63 m.

In the future, the dual-antenna RTK positioning system should be enhanced by fusing inertial navigation system, which not only provides the position and attitude at high update frequency, but also enables precise lever-arm identification and reliable seamless navigation. In addition, the calculation of PSO optimalization is based on the fitness functions (8) and (9), if more tracking deviations are provided, the optimal result would be better, especially when the curvature and environment disturbances change randomly. With the increase in the number of control periods, the computational complexity is increased too, and thus running the optimalization on a FPGA chip to speed up the calculation of PSO is also part of our future work.

Author Contributions: Conceptualization, Y.S. and B.C.; methodology, B.C.; software, Y.S.; validation, Y.S., F.J. and B.C.; formal analysis, Y.Z.; investigation, B.C. and X.W.; resources, B.C.; writing—original draft preparation, Y.S.; writing—review and editing, B.C.; supervision, B.C.; funding acquisition, B.C. All authors have read and agreed to the published version of the manuscript.

Funding: This research was funded by National Natural Science Foundation of China under grant number 31901416, in part by the Primary Research and Development Plan of Jiangsu Province under grant number BE2021313, in part by the Postdoctoral Science Foundation of China under grant 2019M651745, in part by the Jiangsu Province and Education Ministry Co-sponsored Synergistic Innovation Center of Modern Agricultural Equipment and the APC was funded by 31901416.

Institutional Review Board Statement: Not applicable.

Informed Consent Statement: Not applicable.

Data Availability Statement: The data presented in this study are available on request from the corresponding author. The data are not publicly available due to privacy of the subjects involved in the study.

Conflicts of Interest: The authors declare no conflict of interest.

Nomenclature

Symbol	Unit	Description
SM		Stanley model
FSM		Fuzzy Stanley model
PSO		Particle swarm optimization
PSO-FSM		PSO-enhanced fuzzy Stanley model
MAE		Mean absolute error
RMSE		Root mean square error
$\delta(t)$	rad	Expected angle
$\delta_e(t)$	rad	Expected angle due to lateral deviation
$\delta_\theta(t)$	rad	Expected angle due to heading deviation
$e(t)$	m	Lateral deviation
$\theta_e(t)$	rad	Heading deviation
$v(t)$	m/s	Driving speed
$\delta_{PSO}(t)$	rad	Expected angle after PSO

References

1. Backman, J.; Oksanen, T.; Visala, A. Navigation system for agricultural machines: Nonlinear model predictive path tracking. *Comput. Electron. Agric.* **2012**, *82*, 32–43. [CrossRef]
2. Yang, Y.; Li, Y.; Wen, X.; Zhang, G.; Ma, Q.; Cheng, S.; Qi, J.; Xu, L.; Chen, L. An optimal goal point determination algorithm for automatic navigation of agricultural machinery: Improving the tracking accuracy of the pure pursuit algorithm. *Comput. Electron. Agric.* **2022**, *194*, 106760. [CrossRef]
3. Ding, C.; Ding, S.; Wei, X.; Mei, K. Composite SOSM controller for path tracking control of agricultural tractors subject to wheel slip. *ISA Trans.* **2022**, in press. [CrossRef]

4. Li, H.; Gao, F.; Zuo, Z. Research on the agricultural machinery path tracking method based on deep reinforcement learning. *Sci. Program.* **2022**, *2022*, 6385972. [CrossRef]
5. Yu, L.; Yan, X.; Kuang, Z.; Chen, B.; Zhao, Y. Driverless bus path tracking based on fuzzy pure pursuit control with a front axle reference. *Appl. Sci.* **2020**, *10*, 230. [CrossRef]
6. Benoit, T.; Philippe, M.; Christophe, C.; Martinet, P. High accuracy path tracking for vehicle in presence of sliding: Application to farm vehicles automatic guidance for agricultural task. *Auton. Robot.* **2006**, *21*, 79–97.
7. Liu, Z.; Zheng, W.; Wang, N.; Lyu, Z.; Zhang, W. Trajectory trackig control of agricultural vehicles based on disturbance test. *Int. J. Agric. Biol. Eng.* **2020**, *13*, 138–145.
8. Park, M.; Lee, S.; Han, W. Development of steering control system for autonomous vehicle using geometry-based path tracking algorithm. *ETRI J.* **2015**, *37*, 617–625. [CrossRef]
9. Jing, Y.; Liu, G.; Luo, C. Path tracking control with slip compensation of a global navigation satellite system based tractor-scraper land levelling System. *Biosyst. Eng.* **2021**, *212*, 360–377. [CrossRef]
10. Hu, J.; Li, T. Cascaded navigation control for agricultural vehicles tracking straight paths. *Int. J. Agric. Biol. Eng.* **2014**, *7*, 36–44.
11. Zhang, Y.; Wang, L.; Liu, Y. Adaptive neural network-based path tracking control for autonomous combine harvester with input saturation. *Ind. Robot.* **2021**, *48*, 510–522. [CrossRef]
12. Yao, Q.; Tian, Y. A model predictive controller with longitudinal speed compensation for autonomous vehicle path tracking. *Appl. Sci.* **2019**, *9*, 4739. [CrossRef]
13. Li, T. Adaptive sliding mode path tracking control of agricultural wheeled mobile robots. *China Mech. Eng.* **2018**, *29*, 579–584. (In Chinese)
14. Elbanhawi, M.; Simic, M.; Jazar, R. Receding horizon lateral vehicle control for pure pursuit path tracking. *J. Vib. Control.* **2018**, *24*, 619–642. [CrossRef]
15. Thrun, T.; Montemerlo, M.; Dahlkamp, H.; Stavens, D.; Aron, A.; Diebel, J.; Fong, P.; Gale, J.; Halpenny, M.; Hoffmann, G.; et al. Stanley: The robot that won the DARPA grand challenge. *J. Field Robot.* **2006**, *23*, 661–692. [CrossRef]
16. Zhang, H.; Wang, G.; Qin, C.; Lu, Y.; Liu, L.; Gong, J. Agricultural machinery automatic navigation control system based on improved pure tracking model. *Trans. Chin. Soc. Agric. Mach.* **2020**, *51*, 18–25. (In Chinese)
17. Wang, L.; Zhai, Z.; Mao, E.; Zhu, Z. Path tracking control of an autonomous tractor using improved Stanley controller optimized with multiple-population genetic algorithm. *Actuators* **2022**, *11*, 22. [CrossRef]
18. Amer, N.H.; Hudha, K.; Zamzuri, H.; Aparow, V.R.; Abidin, A.F.Z.; Kadir, Z.A.; Murrad, M. Adaptive modified Stanley controller with fuzzy supervisory system for trajectory tracking of an autonomous armoured vehicle. *Robot. Auton. Syst.* **2018**, *105*, 94–111. [CrossRef]
19. Bhattacharyya, B.; Goswami, S.K. Reactive power optimization through evolutionary techniques: A comparative study of the GA, DE and PSO algorithms. *Intell. Autom. Soft Comput.* **2007**, *13*, 461–469. [CrossRef]
20. Khelil, C.; Amrouche, B.; Kara, K.; Chouder, A. The impact of the ANN's choice on PV systems diagnosis quality. *Energy Conv. Manag.* **2021**, *240*, 114278. [CrossRef]
21. Xiong, Z.; Ye, Z.; He, J.; Chen, L.G.; Linghu, J.Q. Small agricultural machinery path intelligent tracking control based on fuzzy immune PID. *Robot.* **2015**, *37*, 212–223. (In Chinese)
22. Ma, J.; Liu, Y.; Zang, S.; Wang, L. Robot path planning based on genetic algorithm fused with continuous Bezier optimization. *Comput. Intell. Neurosci.* **2020**, *2020*, 9813040. [CrossRef] [PubMed]
23. Sharma, P.; Sharma, H.; Kumar, S.; Sharma, K. Black-hole Gbest differential evolution algorithm for solving robot path planning problem. In *Harmony Search and Nature Inspired Optimization Algorithms*; Yadav, N., Yadav, A., Bansal, J.C., Deep, K., Kim, J.H., Eds.; Advances in Intelligent Systems and Computing; Springer: Singapore, 2019; Volume 741, pp. 1009–1022.
24. Tao, Q.; Sang, H.; Guo, H.; Ping, W. Improved particle swarm optimization algorithm for AGV path planning. *IEEE Access* **2021**, *9*, 33522–33531.
25. Sun, H.; Bi, H.; Wang, J.; Zhang, Y.; Wang, Y.; Yu, Z.; Yang, X.; Chai, Y. Prediction of developing modern agriculture demands for the agricultural scientific research institutions services based on BP artificial neural network. In Proceedings of the 2nd International Conference on Biofilms (ChinaBiofilms 2019), Guangzhou, China, 10–13 October 2019.
26. Huang, Q.; Ma, Y.; Zhang, Z. Data preprocessing for agricultural IoT based on RBF neural network. In Proceedings of the 5th Annual International Conference on Network and Information Systems for Computers (ICNISC2019), Wuhan, China, 19–20 April 2019.
27. Guan, S.; Fang, Q.; Guan, T. Application of a novel PNN evaluation algorithm to a greenhouse monitoring system. *IEEE Trans. Instrum. Meas.* **2021**, *70*, 2510712. [CrossRef]
28. Xu, X.; Chen, S.; Ren, L.; Han, C.; Lv, D.; Zhang, Y.; Ai, F. Estimation of heavy metals in agricultural soils using Vis-NIR spectroscopy with fractional-order derivative and generalized regression neural network. *Remote Sens.* **2021**, *13*, 2718. [CrossRef]
29. Nie, J.; Lin, X. Improved adaptive integral line-of-sight guidance law and adaptive fuzzy path following control for underactuated MSV. *ISA Trans.* **2019**, *94*, 151–163. [CrossRef]
30. Guan, Z.; Li, Y.; Mu, S.; Zhang, M.; Jiang, T.; Li, H.; Wang, G.; Wu, C. Tracing algorithm and control strategy for crawler rice combine harvester auxiliary navigation system. *Biosyst. Eng.* **2021**, *211*, 50–62. [CrossRef]
31. Haider, M.J.; Aqeel, M.J.; Rosdiadee, N. Accurate empirical path-loss model based on particle swarm optimization for wireless sensor networks in smart agriculture. *IEEE Sens. J.* **2020**, *20*, 552–561.

32. Evan, K.; Scott, A.; Luis, R. Autonomous surface vehicle energy-efficient and reward-based path planning using particle swarm optimization and visibility graphs. *Appl. Ocean Res.* **2022**, *122*, 103125.
33. Nie, T.; Zhang, Y.; Zhao, Y.; Fang, B.; Zhang, L. Wide-area optimal damping control for power systems based on the ITAE criterion. *Int. J. Electr. Power Energy Syst.* **2019**, *106*, 192–200. [CrossRef]
34. Yin, X.; An, J.; Wang, Y.; Wang, Y.; Jin, C. Development and experiments of the autonomous driving system for high-clearance spraying machines. *Trans. Chin. Soc. Agric. Mach.* **2021**, *37*, 22–30. (In Chinese)

applied sciences

Article

A Novel Decoupled Synchronous Control Method for Multiple Autonomous Unmanned Linear Systems: Bounded L_2-Gain for Coupling Attenuation

Yinsheng Li, Bing Wang * and Yuquan Chen

College of Energy and Electrical Engineering, Hohai University, Nanjing 211100, China; 15150693623@139.com (Y.L.); cyq@mail.ustc.edu.cn (Y.C.)
* Correspondence: icekingking@hhu.edu.cn

Featured Application: This proposed method is suitable for the cooperative consensus control of various homogeneous Multiple Autonomous Unmanned linear systems, such as an underwater robot swarm and aerial UAV swarm.

Abstract: This paper addresses the distributed optimal decoupling synchronous control of multiple autonomous unmanned linear systems (MAUS) subject to complex network dynamic coupling. The leader–follower mechanism based on neighborhood error dynamics is established and the network coupling term is regarded as the external disturbance to realize the decoupling cooperative control of each agent. The Bounded L_2-Gain problem for the network coupling term is formulated into a multi-player zero-sum differential game. It is shown that the solution to the multi-player zero-sum differential game requires the solution to coupled Hamilton–Jacobi (HJ) equations. The coupled HJ equations are transformed into an algebraic Riccati equation (ARE), which can be solved to obtain the Nash equilibrium of a multi-player zero-sum game. It is shown that the bounded L_2-Gain for coupling attenuation can be realized by applying the zero-sum game solution as the control protocol and the ultimately uniform boundedness (UUB) of a local neighborhood error vector under conservative conditions is proved. A simulation example is provided to show the effectiveness of the proposed method.

Keywords: synchronous control; multiple autonomous unmanned linear systems; bounded L_2-gain; multi-player zero-sum game; coupling attenuation; coupled Hamilton–Jacobi equation; nash equilibrium; algebraic Riccati equation

1. Introduction

In the field of multi-agent distributed control, synchronous cooperative control is one of the most popular research topics because of its wide application prospects in many engineering systems, such as the cooperative control of autonomous underwater vehicles, wind farm and unmanned aerial vehicles. A great deal of research has been developed on the distributed control methods for multi-agent synchronization [1–4]. The purpose of distributed synchronous control is to design a control protocol for each agent, depending only on the states of neighboring agents, to ensure that the states of all agents in the communication digraph finally achieve synchronization. A leader–follower mechanism is the most popular one for distributed synchronous control because of its simplicity and scalability. The basic idea is that a leader agent is set as the reference node, and other agents are set as the follower nodes whose goal is to track the reference node to achieve the ultimate synchronization of the entire communication network.

On the basis of this mechanism, ref. [5] defines the local neighborhood error of each agent and deduces that this is a dynamical system with multiple control inputs, from itself and all of its neighbors. This means that the local neighborhood error of each agent is the result of coupling from adjacent node agents, which brings considerable complexity to the design of the control protocol. Ref. [6] formulated this intricate relationship as a differential game, namely, a multi-agent differential graphical game, by defining a local performance index for each agent. Optimal control and game theory [7,8] has been successful utilized to formulate strategic behavior, where the dynamic of each agent relies on the actions of itself and its neighbors. In an optimal control and differential graphical game, each agent minimizes the performance objective (cost value function) by adjusting its control strategy to optimal. In [9], the finite-time optimal coordination problem of multi-agent systems (MASs) is investigated. The authors in [10] addressed the centralized optimal coordination problem under tree formation constraints. In [11], the robust optimal formation control of heterogeneous MASs is studied. These published graphical games and optimal control methods for consensus and synchronization are achieved based on the solution of coupled Hamilton–Jacobi–Isaacs (HJI) equations and Hamilton–Jacobi–Bellman (HJB) equations, respectively. In practice, coupled HJI and coupled HJB equation are difficult to be solved by analytical methods due to their inherent nonlinearity and uncertainty.

The reinforcement learning (RL) method is often regarded as the effective method to solve the coupled HJI and coupled HJB equation. RL is the branch of machine learning concerned with how to methodically adjust the control strategy of agents based on the rewards from the environment [12–15]. In [16], an online distributed optimal adaptive algorithm is proposed for a differential graphical game, the intelligent identifier is designed to find the unknown dynamic and the neural actor–critic network structure is introduced to find the solutions of the Nash equilibrium. In [17], the bounded L_2-gain consensus problem for the MASs with external disturbance is formulated into the zero-sum differential game by introducing a specific performance index and a policy iteration (PI) algorithm-based RL is provided to find the solution to the coupled HJI equations. In [18], the optimal synchronization control problem is studied for homogeneous MASs with input saturation by using the RL methods. This research utilizes the neural network as an approximator and a design-specific update law so that the neural network approximates the optimal value function and optimal control strategy with certain precision. However, the strict asymptotic convergence proof of the neural network is not given in these works, and only the boundedness of approximate errors is guaranteed. In addition, the neural network approximator-based RL needs to satisfy the persistence of the excitation condition (PE) [19–23], which also limits the practical engineering application of these methods.

The quadratic optimal control problem of a linear single system can be solved by solving an algebraic Riccati equation (ARE) [24], but the optimal control problem of MASs is far more complicated than that of a single system owing to the state coupling in the control design. At present, some optimal control methods of MASs are accompanied by a huge amount of calculations and strong assumptions.

Motivated by the above discussion, this paper focuses on the optimal cooperative control of Multiple Autonomous Unmanned linear systems (MAUS) from a new perspective, i.e., the adjacent nodes' input coupling part is regarded as the external disturbance. Thus, the complex distributed multi-agent error dynamics are decoupled into centralized multi-input dynamics. Inspired by the idea of a zero-sum game in [17], this paper formulates these centralized multi-input dynamics into multiple independent multi-player zero-sum differential games. The motivation is to realize the decoupled optimal synchronous control of MASs and the main contributions of this paper are listed in the following:

(1) The coupling among the distributed multi-agents is equivalent to the disturbance from different channels, and the local neighborhood error dynamics of each agent are modeled as an independent centralized multi-player game.
(2) The bounded L_2-gain problem for coupling attenuation is introduced and is formulated into a multi-player zero-sum game by defining a modified performance index. Different from the L_2-gain problem of [17], concerning disturbance rejection, the motive of the bounded L_2-gain for the coupling attenuation studied here is to suppress the coupling effect on the performance.

(3) It is proved that the solution of the zero-sum game requires the solution of the coupled Hamilton–Jacobi (HJ) equation. The coupled HJ equation of each agent is transformed into an independent equivalent algebraic Riccati equation, which simplifies the solution process effectively.

This paper is organized as follows. Section 2 provides the mathematical background and derives the local error dynamics of each node that is coupled by its own control protocol and those of its neighbors. Section 3 proposes the problem formulation of the bounded L_2-gain for coupling attenuation and its equivalent multi-player zero-sum differential game. Section 4 transforms this zero-sum differential game into the solution of an algebraic Riccati equation and proves the ultimately uniform boundedness of the local neighborhood error, conservatively. The simulation results and conclusion is presented in Sections 5 and 6, respectively.

2. Preliminaries and Problem Formulation

2.1. Graph Theory

In this paper, the multi-agent directed communication network is depicted. A directed connected graph is defined as $G(V, E, A)$, where $V = \{v_1, v_2, \cdots, v_n\}$ represents a finite non-empty set of nodes, $E \subseteq V \times V$ is the ordered set of nodes pairs and $A = [a_{ij}]$ is the adjacency matrix. If node v_i can receive the information from node v_j, then the node pairs $\overline{v}_{ij} = (v_i, v_j) \in E$, and node v_j is called a neighbor of node v_i. The neighbor set of node v_i is represented by $N_i = \{ v_j | (v_i, v_j) \in E \}$. Correspondingly, the adjacency matrix element $a_{ij} = 1$ when $v_j \in N_i$, otherwise $a_{ij} = 0$. The graph Laplacian matrix is defined as $L = D - A$, whose row sums are equal to zero [25]. Diagonal matrix $D = diag(d_i)$ is the in-degree matrix, where $d_i = \sum_{j \in N_i} a_{ij}$ is the in-degree of node v_i.

Definition 1. *A directed graph is called as strongly connected if there is a directed path for any a pair of distinct nodes* (v_i, v_j), *where the directed path is the edge sequence* $(v_{i1}, v_{i2}), (v_{i2}, v_{i3}), \ldots, (v_{ik}, v_j)$.

Definition 2 [17]. *A directed tree is a connected graph where every node except the root node, has an in-degree equal to one. The graph is called to have a spanning tree if a subset of the node pairs constructs a directed tree.*

In this paper, $\overline{\lambda}(B)$ and $\underline{\lambda}(B)$ represent the maximum and minimum singular values of the matrix B, respectively.

2.2. Problem Formulation

Considering the Multiple Autonomous Unmanned linear systems (MAUS) constructed by the directed communication graph $G(V, E, A)$ having N agents, the dynamics of each agent is described in the following:

$$\dot{x}_i = Ax_i + Bu_i \tag{1}$$

where $x_i \in R^n$, $u_i \in R^{m_i}$ are states and control inputs of node i, respectively. The cooperative control of homogeneous systems is investigated in this paper and the leader node $x_0 \in R^n$ is set to satisfy the following dynamic

$$\dot{x}_0 = Ax_0 \tag{2}$$

The problem of MAUS synchronization is designing control protocols u_i for each agent so that states of each node track the leader node, i.e., $\|x_0 - x_i\| \to 0, \forall i$.

The neighborhood error for each node is defined as [26]

$$\delta_i = \sum_{j \in N_i} a_{ij}(x_i - x_j) + g_i(x_i - x_0) \tag{3}$$

where $g_i \geq 0$ denotes the pinning gain and there is at least one node that has a link to the leader node.

For the neighborhood error (3), the overall neighborhood error vector of graph $G(V, E, A)$ is given by

$$\delta = ((L + \overline{G}) \otimes I_n)(x_i - \underline{x}_0) \tag{4}$$

where $x = [x_1^T x_2^T \cdots x_N^T]^T$ and $\delta = [\delta_1^T \delta_2^T \cdots \delta_N^T]^T$ denote the global state vector and global error vector, respectively. Moreover, for $\underline{x}_0 = \underline{I}x_0 \in R^{nN}$ with $\underline{I} = 1_N \otimes I_n \in R^{nN \times n}$, I_n denotes the n dimensional identity matrix and 1_N denotes the N-vector of ones. The symbol $\otimes$ is the Kronecker product [27]. $\overline{G} = diag(g_1, g_2, \cdots, g_N)$ as a diagonal matrix represents the connection between all agents and the leader node.

The overall synchronization error is

$$\varepsilon = (x - \underline{x}_0) \in R^{nN} \tag{5}$$

Assumption 1. *The communication graph is strongly connected, i.e., there is a directed path for any a pair of distinct nodes.*

On the basis of Assumption 1, if $\overline{G} \neq 0$, then $g_i \neq 0$ for at least one. In this case, the matrix $L + \overline{G}$ is non-singular and the real parts of all eigenvalues are positive [26]. The following lemma can be obtained, which shows that the overall neighborhood error vector δ is positively correlated with the overall synchronization error ε.

Lemma 1. *If the communication graph is strongly connected and $\overline{G} \neq 0$, the synchronization errors are bounded, as follows*

$$\delta / \overline{\lambda}(L + \overline{G}) \leq \|\varepsilon\| \leq \delta / \underline{\lambda}(L + \overline{G}) \tag{6}$$

Furthermore, $\delta \equiv 0$ if and only if all nodes are synchronized, i.e.,

$$\varepsilon = (x - \underline{x}_0) = 0 \tag{7}$$

The dynamics of the local neighborhood tracking errors are given as

$$\dot{\delta}_i = \sum_{j \in N_i} a_{ij}(\dot{x}_i - \dot{x}_j) + g_i(\dot{x}_i - \dot{x}_0) \tag{8}$$

Substituting (1) and (2) into the above equation, it can be obtained that

$$\dot{\delta}_i = A\delta_i + (d_i + g_i)B_i u_i - \sum_{j \in N_i} a_{ij} B_j u_j \tag{9}$$

It can be seen that the dynamics of the local neighborhood error of each agent i is affected by multiple control inputs from node i and its adjacent nodes. The whole MAUS with the communication graph $G(V, E, A)$ presents a complex coupling relationship. It is quite intricate to solve the optimal control problem of dynamic (9) affected by multi-coupling.

3. Multi-Player Zero-Sum Differential Game for Decoupled Multi-Agent System

3.1. The Bounded L_2-Gain Problem for Coupling Attenuation of Multi-Agent System

For decoupling, the inputs from adjacent nodes in the dynamics (9) are replaced by the virtual coupling actions $w_i(t) = [\widehat{u}_j^{T}]^T_{j \in N_j}$ which is regarded as the external disturbances. The performance output is defined as $z_i(t) = [\delta_i^T u_i^T]^T$. It is desired to designed the control protocol u_i to achieve synchronization while satisfying the follow bounded L_2-gain condition for the coupling actions with a given $\gamma_i > 0$

$$\begin{aligned} \int_0^T \|z_i(t)\|^2 dt &= \int_0^T (\delta_i^T Q_i \delta_i + u_i^T R_{ii} u_i)\, dt \\ &\le \gamma_i^2 \int_0^T \sum_{j\in N_i} \widehat{u}_j^T R_{ij} \widehat{u}_j\, dt + \beta(\delta_i(0)) \end{aligned} \tag{10}$$

where, $\beta(\cdot)$ is a bounded function such that $\beta(0) = 0, Q_i > 0, R_{ii} > 0, R_{ij} > 0$. γ_i^* is defined as the minimum value of γ_i while the bounded L_2-gain condition (10) is satisfied.

3.2. Multi-Player Zero-Sum Differential Game

The following equation is used to define the following performance index function for each agent.

$$J_i(\delta_i(0), u_i, \widehat{u}_{-i}) = \frac{1}{2}\int_0^\infty \delta_i^T Q_i \delta_i + u_i^T R_{ii} u_i - \gamma_i^2 \sum_{j\in N_i} \widehat{u}_j^T R_{ij} \widehat{u}_j dt \tag{11}$$

where $\widehat{u}_{-i}$ denotes the virtual coupling control inputs from neighboring nodes, i.e., $\widehat{u}_{-i} = \left\{ \widehat{u}_j | \, j \in N_i \right\}$. It should be noted that the main difference from [17] is that the coupling control inputs from neighboring nodes are regarded as the virtual external disturbances directly, which greatly simplifies the design of the control protocol u_i.

The solution for the bounded L_2-gain problem for coupling attenuation depicted in Section 3.1 can be equivalent to the Nash equilibrium solution of the multi-player zero-sum game-base on the performance index function (11). That is

$$V_i(\delta_i(0)) = \min_{u_i}\max_{\widehat{u}_{-i}} J_i(\delta_i(0), u_i, \widehat{u}_{-i}) \tag{12}$$

In this multi-player zero-sum game, the goal of u_i is to minimize the value $V_i(\delta_i(0))$. On the contrary, the virtual coupling inputs $\widehat{u}_{-i}$ are assumed to maximize the value. This game has a unique solution if a game-theoretic saddle point $(u_i^*, \widehat{u}_{-i}^*)$ exists, i.e.,

$$V_i^*(\delta_i(0)) = \min_{u_i}\max_{\widehat{u}_{-i}} J_i(\delta_i(0), u_i, \widehat{u}_{-i}) = \max_{\widehat{u}_{-i}}\min_{u_i} J_i(\delta_i(0), u_i, \widehat{u}_{-i}) \tag{13}$$

Accordingly, the value $V_i^*(\delta_i(0))$ in the above equation is the value of the zero-sum game and satisfies the following Nash equilibrium condition for all policies $u_i, \widehat{u}_{-i}$

$$J_i(\delta_i(0), u_i^*, \widehat{u}_{-i}) \le J_i(\delta_i(0), u_i, \widehat{u}_{-i}) \le J_i(\delta_i(0), u_i, \widehat{u}_{-i}^*) \tag{14}$$

When the policies $u_i, \widehat{u}_{-i}$ are selected, the value function of node i can yield

$$V_i(\delta_i(t), u_i, \widehat{u}_{-i}) = \frac{1}{2}\int_t^\infty \delta_i^T Q_i \delta_i + u_i^T R_{ii} u_i - \gamma_i^2 \sum_{j\in N_i} \widehat{u}_j^T R_{ij} \widehat{u}_j dt \tag{15}$$

Differential equivalents to each value function are given as

$$\begin{aligned} 0 = {} & \tfrac{1}{2}\left(\delta_i^T Q_i \delta_i + u_i^T R_{ii} u_i - \gamma_i^2 \sum_{j\in N_i} \widehat{u}_j^T R_{ij} \widehat{u}_j \right) \\ & + \nabla V_i^T (A\delta_i + (d_i + g_i) B_i u_i - \sum_{j\in N_i} a_{ij} B_j \widehat{u}_j), \; V_i(0) = 0, \; i \in N \end{aligned} \tag{16}$$

where $\nabla V_i = \frac{\partial V_i}{\partial \delta_i} \in R^n$ denotes the gradient vector. The Hamiltonian functions are defined as follows,

$$\begin{aligned} H_i(\delta_i, \nabla V_i, u_i, \widehat{u}_{-i}) \equiv \frac{1}{2}\left(\delta_i^T Q_i \delta_i + u_i^T R_{ii} u_i - \gamma_i^2 \sum_{j \in N_i} \widehat{u}_j^T R_{ij} \widehat{u}_j \right) \\ + \nabla V_i^T (A\delta_i + (d_i + g_i) B_i u_i - \sum_{j \in N_i} a_{ij} B_j \widehat{u}_j) \end{aligned} \tag{17}$$

Under certain policies u_i, $\widehat{u}_{-i}$, the partial differential equation $H_i(\delta_i, \nabla V_i, u_i, \widehat{u}_{-i}) = 0$ has a unique solution $V_i(\delta_i)$. The principle of optimality gives

$$\begin{aligned} \frac{\partial H_i(\delta_i, \nabla V_i, u_i, \widehat{u}_{-i})}{\partial u_i} = 0 \Rightarrow u_i = -(d_i + g_i) R_{ii}^{-1} B_i^T \nabla V_i \\ \frac{\partial H_i(\delta_i, \nabla V_i, u_i, \widehat{u}_{-i})}{\partial \widehat{u}_j} = 0 \Rightarrow \widehat{u}_j = -\frac{1}{\gamma_i^2} a_{ij} R_{ij}^{-1} B_j^T \nabla V_i,\ j \in N_i \end{aligned} \tag{18}$$

If the V_i^* is the Nash equilibrium solution of the multi-player zero-sum game, that is

$$V_i^*(\delta_i) = \min_{u_i} \max_{\widehat{u}_{-i}} J_i(\delta_i, u_i, \widehat{u}_{-i}) \tag{19}$$

we can obtain

$$\min_{u_i} \max_{\widehat{u}_{-i}} H_i(\delta_i, \nabla V_i^*, u_i, \widehat{u}_{-i}) = 0 \tag{20}$$

Substituting the optimal strategy determined by (18) into (20), the coupled Hamilton–Jacobi (HJ) equations yield

$$\begin{aligned} 0 = \frac{1}{2}\delta_i^T Q_i \delta_i + \frac{1}{2}(d_i + g_i)^2 \nabla V_i^T B_i R_{ii}^{-1} B_i^T \nabla V_i - \frac{1}{2\gamma_i^2} \sum_{j \in N_i} a_{ij}^2 \nabla V_i^T B_j R_{ij}^{-1} B_j^T \nabla V_i \\ + \nabla V_i^T (A\delta_i - (d_i + g_i)^2 B_i R_{ii}^{-1} B_i^T \nabla V_i + \frac{1}{\gamma_i^2} \sum_{j \in N_i} a_{ij}^2 B_j R_{ij}^{-1} B_j^T \nabla V_i),\ V_i(0) = 0,\ i \in N \end{aligned} \tag{21}$$

For a given solution V_i^*, in order to define $u_i^* = u_i(V_i^*)$ and $\widehat{u}_j^* = \widehat{u}_j(V_i^*)$ in the same way as (18), (21) can be written as

$$H_i(\delta_i, \nabla V_i^*, u_i^*, \widehat{u}_{-i}^*) = 0,\ V_i^*(0) = 0 \tag{22}$$

Lemma 2. *For any policies u_i, $\widehat{u}_{-i}$, the following equation holds*

$$H_i(\delta_i, \nabla V_i^*, u_i, \widehat{u}_{-i}) = \frac{1}{2}(u_i - u_i^*)^T R_{ii} (u_i - u_i^*) - \sum_{j \in N_i} \frac{\gamma_i}{2} (\widehat{u}_j - \widehat{u}_j^*)^T R_{ij} (\widehat{u}_j - \widehat{u}_j^*) \tag{23}$$

Proof of Lemma 2. Substituting u_i, $\widehat{u}_{-i}$ for u_i^*, $\widehat{u}_{-i}^*$ in (22)

$$\begin{aligned} H_i(\delta_i, \nabla V_i^*, u_i, \widehat{u}_{-i}) = \nabla V_i^{*T} (A\delta_i + (d_i + g_i) B_i u_i - \sum_{j \in N_i} a_{ij} B_j \widehat{u}_j \\ + (d_i + g_i) B_i u_i^* - (d_i + g_i) B_i u_i^* - \sum_{j \in N_i} a_{ij} B_j \widehat{u}_j^* + \sum_{j \in N_i} a_{ij} B_j \widehat{u}_j^*) \\ + \frac{1}{2}\left(\delta_i^T Q_i \delta_i + u_i^T R_{ii} u_i - \gamma_i^2 \sum_{j \in N_i} \widehat{u}_j^T R_{ij} \widehat{u}_j \right) \\ + \frac{1}{2}\left(u_i^{*T} R_{ii} u_i^* - u_i^{*T} R_{ii} u_i^* - \gamma_i^2 \sum_{j \in N_i} \widehat{u}_j^{*T} R_{ij} \widehat{u}_j^* + \gamma_i^2 \sum_{j \in N_i} \widehat{u}_j^{*T} R_{ij} \widehat{u}_j^* \right) \end{aligned} \tag{24}$$

Substituting $H_i(\delta_i, \nabla V_i^*, u_i^*, \widehat{u}_{-i}^*) = 0$ into (24), we can obtain

$$\begin{aligned} H_i(\delta_i, \nabla V_i^*, u_i, \widehat{u}_{-i}) &= \nabla V_i^{*T}((d_i+g_i)B_i(u_i-u_i^*) - \sum_{j\in N_i} a_{ij}B_j(\widehat{u}_j - \widehat{u}_j^*)) \\ &+\tfrac{1}{2}(u_i^T R_{ii}u_i - u_i^{*T}R_{ii}u_i^*) - \tfrac{1}{2}\left(\gamma_i^2 \sum_{j\in N_i} \widehat{u}_j^T R_{ij}\widehat{u}_j - \gamma_i^2 \sum_{j\in N_i} \widehat{u}_j^{*T} R_{ij}\widehat{u}_j^*\right) \end{aligned} \tag{25}$$

Completing the squares in (25) gives (23) upon the relationship between $u_i^*, \widehat{u}_{-i}^*$ and ∇V_i^*. □

Remark 1. *The polices $\widehat{u}_{-i}$ in Section 3 are not the real policies of neighboring nodes. These are only defined as the virtual coupling input from neighboring nodes, which are regarded as the external disturbances and have the same channels as the control inputs of neighboring nodes. In this way, the bounded L_2-gain attenuation for the real coupling inputs from neighboring nodes can be realized. In addition, the complex relationships among agents are decoupled virtually during the control protocol design process, and the solving process of zero-sum game is effectively simplified. The coupled HJ equation of each agent is independent of each other.*

4. Solution to Bounded L_2-Gain Problem for the Coupling Attenuation and the Equivalent Algebraic Riccati Equation

4.1. Solution to Bounded L_2-Gain Problem for Coupling Attenuation

In this subsection, the control policy u_i is found to guarantee the condition (10) holds for a prescribed $\gamma_i > 0$ and $\widehat{u}_{-i} \in L_2[0,\infty)$. The following Theorem 1 shows that the solution of a coupled HJ equation (22) is actually the solution to the bounded L_2-gain problem for coupling attenuation.

Theorem 1. *Let $\gamma_i \geq \gamma_i^*$. Suppose the coupled HJ equation (22) has a smooth positive definite solution $V_i^* > 0, i \in N$. The control policy is selected as $u_i^* = u_i(V_i^*)$, given by (18) in terms of V_i^*. The bounded L_2-gain condition (10) holds for all $\widehat{u}_{-i} \in L_2[0,\infty)$.*

Proof of Theorem 1. According to Lemma 2

$$\begin{aligned} &H_i(\delta_i, \nabla V_i^*, u_i, \widehat{u}_{-i}) \\ &= \tfrac{1}{2}\left(\delta_i^T Q_i\delta_i + u_i^T R_{ii}u_i - \gamma_i^2 \sum_{j\in N_i} \widehat{u}_j^T R_{ij}\widehat{u}_j\right) + \tfrac{dV_i^*}{dt} \\ &= \tfrac{1}{2}(u_i - u_i^*)^T R_{ii}(u_i - u_i^*) - \sum_{j\in N_i} \tfrac{\gamma_i}{2}(\widehat{u}_j - \widehat{u}_j^*)^T R_{ij}(\widehat{u}_j - \widehat{u}_j^*) \end{aligned} \tag{26}$$

Selecting $u_i = u_i^*$, we can obtain that

$$\begin{aligned} &\tfrac{1}{2}\left(\delta_i^T Q_i\delta_i + u_i^T R_{ii}u_i - \gamma_i^2 \sum_{j\in N_i} \widehat{u}_j^T R_{ij}\widehat{u}_j\right) + \tfrac{dV_i^*}{dt} \\ &= -\sum_{j\in N_i} \tfrac{\gamma_i}{2}(\widehat{u}_j - \widehat{u}_j^*)^T R_{ii}(\widehat{u}_j - \widehat{u}_j^*) \leq 0 \end{aligned} \tag{27}$$

Integrating (27) yields,

$$\begin{aligned} &\tfrac{1}{2}\int_0^T \left(\delta_i^T Q_i\delta_i + u_i^T R_{ii}u_i - \gamma_i^2 \sum_{j\in N_i} \widehat{u}_j^T R_{ij}\widehat{u}_j\right) dt \\ &+V_i^*(\delta_i(T)) - V_i^*(\delta_i(0)) \leq 0 \end{aligned} \tag{28}$$

V_i^* is a smooth positive definite solution, i.e., $V_i^*(\delta_i(T)) \geq 0$, one has

$$\int_0^T \left(\delta_i^T Q_i \delta_i + u_i^T R_{ii} u_i\right) dt \leq \gamma_i^2 \int_0^T \sum_{j \in N_i} \widehat{u}_j^T R_{ij} \widehat{u}_j \, dt + V_i^*(\delta_i(0)) \tag{29}$$

Hence, the bounded L_2-gain condition (10) for coupling attenuation is satisfied. □

4.2. The Equivalent Algebraic Riccati Equation

It can be seen from the above results that the Nash equilibrium solution can be obtained by solving the coupled HJ equation (22). In this subsection, it will be shown that the coupled HJ equation (22) can be equivalent to an Algebraic Riccati equation (ARE).

Defining the optimal value function $V_i^* = \delta_i^T P_i \delta_i$, the corresponding optimal u_i^* and $\widehat{u}_{-i}^*$ can be obtained as

$$\begin{aligned} u_i^* &= -2(d_i + g_i) R_{ii}^{-1} B_i^T P_i \delta_i \\ \widehat{u}_j^* &= -\tfrac{2}{\gamma_i^2} a_{ij} R_{ij}^{-1} B_j^T P_i \delta_i \, , \ j \in N_i \end{aligned} \tag{30}$$

Substituting (30) and $V_i^* = \delta_i^T P \delta_i$ into (21) yields

$$\begin{aligned} \delta_i^T (P_i A + A^T P_i) \delta_i + \tfrac{1}{2} \delta_i^T Q_i \delta_i \ & -2(d_i + g_i)^2 \delta_i^T P_i B_i R_{ii}^{-1} B_i^T P_i \delta_i \\ & + \tfrac{2}{\gamma_{ij}^2} \delta_i^T \Big(\sum_{j \in N_i} a_{ij}^2 P_i B_j R_{ij}^{-1} B_j^T P_i \Big) \delta_i = 0 \end{aligned} \tag{31}$$

The above equation can be equivalent to

$$\begin{aligned} P_i A + A^T P_i + \tfrac{1}{2} Q_i - 2(d_i + g_i)^2 P_i B_i R_{ii}^{-1} B_i^T P_i \\ + \tfrac{2}{\gamma_{ij}^2} \sum_{j \in N_i} a_{ij}^2 P_i B_j R_{ij}^{-1} B_j^T P_i = 0 \end{aligned} \tag{32}$$

Defining the integrated matrix as

$$\begin{aligned} R &= diag\Big(\tfrac{R_{ii}}{2(d_i+g_i)^2}, -\tfrac{\gamma_i^2 R_{ij_1}}{2a_{ij_1}}, -\tfrac{\gamma_i^2 R_{ij_2}}{2a_{ij_2}}, \cdots, -\tfrac{\gamma_i^2 R_{ij_{d_i}}}{2a_{ij_{d_i}}}\Big) \\ B &= [B_i B_{j_1} B_{j_2} \cdots B_{j_{d_i}}], \ j_1, \ j_2, \ \cdots, j_{d_i} \in N_i \end{aligned}$$

Then, (32) can be rewritten as the ARE

$$P_i A + A^T P_i + \frac{1}{2} Q_i - P_i B R^{-1} B^T P_i = 0 \tag{33}$$

Theorem 2. *Assume that all the real control policies $u_{-i} = \{ u_j | \, j \in N_i \}$ of neighboring agents satisfy the bounded condition as follows,*

$$\|u_j\|_2 = u_j^T u_j \leq \xi, \ j \in N_i \tag{34}$$

Selecting the first equation of (30) as the control policies u_i, based on the solution of ARE (34). The local neighborhood error vector will ultimately and uniformly enter the following bounded invariant set

$$\Omega_i = \{ \, \delta_i | \|\delta_i\|_2 \leq \frac{\gamma_i^2 \xi \sum_{j \in N_i} \overline{\lambda}(R_{ij})}{2 \underline{\lambda}(\overline{Q}_i)} \} \tag{35}$$

where $\|\delta_i\|_2$ denotes the Euclidean norm of δ_i and $\overline{Q}_i$ is the positive definite matrix as $\overline{Q}_i = \frac{1}{2} Q_i + 2(d_i + g_i)^2 P_i B_i R_{ii}^{-1} B_i^T P_i$.

Proof of Theorem 2. Selecting the optimal positive value function $V_i^* = \delta_i^T P_i \delta_i \geq 0$, in terms of the solution of the coupled HJ equation (21), as the Lyapunov function. According to Lemma 2, the derivative of V_i^* is

$$\begin{aligned}\frac{dV_i^*}{dt} = &-\frac{1}{2}\left(\delta_i^T Q_i \delta_i + u_i^T R_{ii} u_i - \gamma_i^2 \sum_{j \in N_i} u_j^T R_{ij} u_j\right) \\ &+\frac{1}{2}(u_i - u_i^*)^T R_{ii}(u_i - u_i^*) - \sum_{j \in N_i} \frac{\gamma_i}{2}\left(u_j - \widehat{u}_j^*\right)^T R_{ij}\left(u_j - \widehat{u}_j^*\right)\end{aligned} \tag{36}$$

Selecting $u_i = u_i^*$ as the first equation of (30) yields

$$\begin{aligned}\frac{dV_i^*}{dt} &= -\frac{1}{2}\delta_i^T Q_i \delta_i - 2(d_i + g_i)^2 \delta_i^T P_i B_i R_{ii}^{-1} B_i^T P_i \delta_i \\ &\quad +\frac{\gamma_i^2}{2} \sum_{j \in N_i} u_j^T R_{ij} u_j - \sum_{j \in N_i} \frac{\gamma_i}{2}\left(u_j - \widehat{u}_j^*\right)^T R_{ij}\left(u_j - \widehat{u}_j^*\right) \\ &\leq -\delta_i^T\left(\frac{1}{2}Q_i + 2(d_i + g_i)^2 P_i B_i R_{ii}^{-1} B_i^T P_i\right)\delta_i + \frac{\gamma_i^2 \xi \sum_{j \in N_i} \overline{\lambda}(R_{ij})}{2}\end{aligned} \tag{37}$$

Let $\frac{dV_i^*}{dt} \leq 0$, we can realize that

$$\delta_i^T \delta_i = \|\delta_i\|_2 \geq \frac{\gamma_i^2 \xi \sum_{j \in N_i} \overline{\lambda}(R_{ij})}{2\underline{\lambda}\left(\frac{1}{2}Q_i + 2(d_i + g_i)^2 P_i B_i R_{ii}^{-1} B_i^T P_i\right)} \tag{38}$$

Defining $\overline{Q}_i = \frac{1}{2}Q_i + 2(d_i + g_i)^2 P_i B_i R_{ii}^{-1} B_i^T P_i$ yields the bounded invariant set (35). □

Remark 2. *Theorem 2 shows the ultimately uniform boundedness (UUB) of the local neighborhood error vector δ_i. According to the bounded invariant sets Ω_i, the bound of δ_i can be arbitrarily small by presetting the matrix Q_i, R_{ii} and R_{ij} in the performance index function (11). In fact, this result is conservative because the term $-\sum_{j \in N_i} \frac{\gamma_i}{2}\left(u_j - \widehat{u}_j^*\right)^T R_{ij}\left(u_j - \widehat{u}_j^*\right)$ in $\frac{dV_i^*}{dt}$ is omitted. The real control inputs u_j of adjacent agents differ greatly from $\widehat{u}_j^*$ in fact, which guarantees the negative characterization of $\frac{dV_i^*}{dt}$. Therefore, the simulation results in the next section show that the local neighborhood error vector can converge asymptotically and uniformly to the origin.*

Remark 3. *In a practical application, matrix Q_i, R_{ii} and R_{ij} and parameter γ_i can be selected according to engineering performance requirements. If a high convergence speed and synchronization accuracy are required, Q_i can be selected to make its eigenvalues large; if a low control energy consumption is required, R_{ii} can be selected to make its eigenvalues large; the coupling attenuation level can be adjusted by adjusting matrix R_{ij} and γ_i. It should be noted that, Q_i, R_{ii}, R_{ij} and γ_i must satisfy the condition in Theorem 1, so that the coupled HJ equation (22) has a smooth positive definite solution $V_i^* > 0$. That is, (33) has a positive definite solution P_i.*

5. Simulation Results

This section shows the effectiveness of the equivalent ARE approach described in Section 4 and Theorem 2. The simulation is realized in MATLAB/Simulink. Consider a class of Multiple Autonomous Unmanned homogeneous linear systems referring to [5] which is shown as follows

$$\dot{x}_i = Ax_i + B_i u_i$$

where $A = \begin{bmatrix} 0 & 1 \\ -1 & 0 \end{bmatrix}$, $B_1 = \begin{bmatrix} 2 \\ 1 \end{bmatrix}$, $B_2 = \begin{bmatrix} 2 \\ 3 \end{bmatrix}$, $B_3 = \begin{bmatrix} 2 \\ 2 \end{bmatrix}$ and $B_3 = \begin{bmatrix} 1 \\ 1 \end{bmatrix}$, with the Leader dynamics $\dot{x}_0 = Ax_0$. The communication digraph structure is shown in Figure 1. The edge weights and the pinning gains are taken equal to 1.

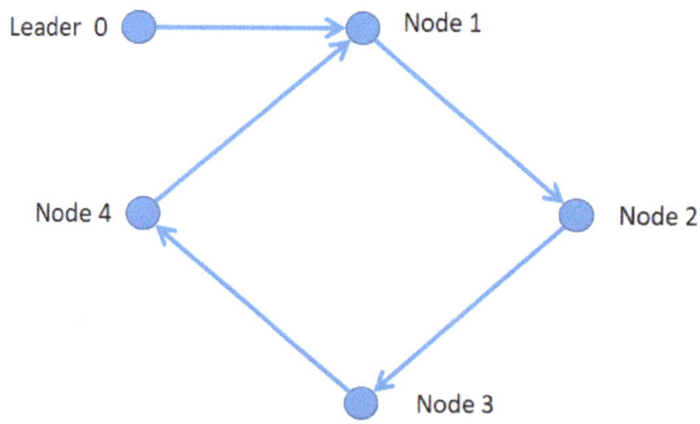

Figure 1. The communication digraph structure.

The selected the weight matrices in (11) are $R_{11} = 9$, $R_{14} = 1$, $R_{22} = 9$, $R_{21} = 1$, $R_{33} = 9$, $R_{32} = 1$, $R_{44} = 9$, $R_{43} = 1$ and $Q_1 = Q_2 = Q_3 = Q_4 = \begin{bmatrix} 8 & 0 \\ 0 & 8 \end{bmatrix}$. The bounded L_2-gain coefficient in (11) for each agent are preset as $\gamma_1 = 1.75$, $\gamma_2 = 3.75$, $\gamma_3 = 4.5$ and $\gamma_4 = 6.25$. The cooperative control protocol of each agent is implemented, as in Section 4.2, where the solution of ARE (33) is

$$P_1 = \begin{bmatrix} 4.8954 & -5.6084 \\ -5.6084 & 9.4934 \end{bmatrix}, P_2 = \begin{bmatrix} 41.9228 & -7.3129 \\ -7.3129 & 2.6742 \end{bmatrix}, P_3 = \begin{bmatrix} 14.4257 & -9.363 \\ -9.363 & 10.4565 \end{bmatrix}, P_4 = \begin{bmatrix} 17.4696 & -0.5066 \\ -0.5066 & 13.6 \end{bmatrix}.$$

Remark 4. *In the process of a simulation design, the γ_i should be gradually reduced to search for a feasible and high coupling attenuation level under the premise that (33) has a positive definite solution P_i. Using the ARE solver in MATLAB, it is very convenient to solve (33) and obtain P_i. Then, the design of the coupling attenuation controller can be completed according to (30).*

To elevate the Bounded L_2-gain problem for the coupling attenuation, the following variable C_γ is introduced based on (29)

$$C_\gamma = \int_0^T \left(\delta_i^T Q_i \delta_i + u_i^T R_{ii} u_i\right) dt - \gamma_i^2 \int_0^T \sum_{j \in N_i} u_j^T R_{ij} u_j \, dt - V_i^*(\delta_i(0)) \quad (39)$$

That is, $C_\gamma \leq 0$, which means that the Bounded L_2-gain condition (10) is satisfied.

The local neighborhood error vector of each agent is shown in Figure 2. Figure 3 is the 3-D phase plane plot of the system's evolution for agents 1, 2, 3, 4 and leader 0. The C_γ of the node 1 agent is shown in Figure 4. As can be seen from Figures 2 and 3, the neighborhood error vector can converge asymptotically and uniformly to the origin and all agents in the communication digraph are eventually synchronized, which is also consistent with Remark 2. Figure 4 shows that C_γ is always negative, which is equivalent to that the node 1 agent satisfies the bounded L_2-gain condition with $\gamma_1 = 1.75$ for the coupling attenuation. The effectiveness of the proposed method is thus verified.

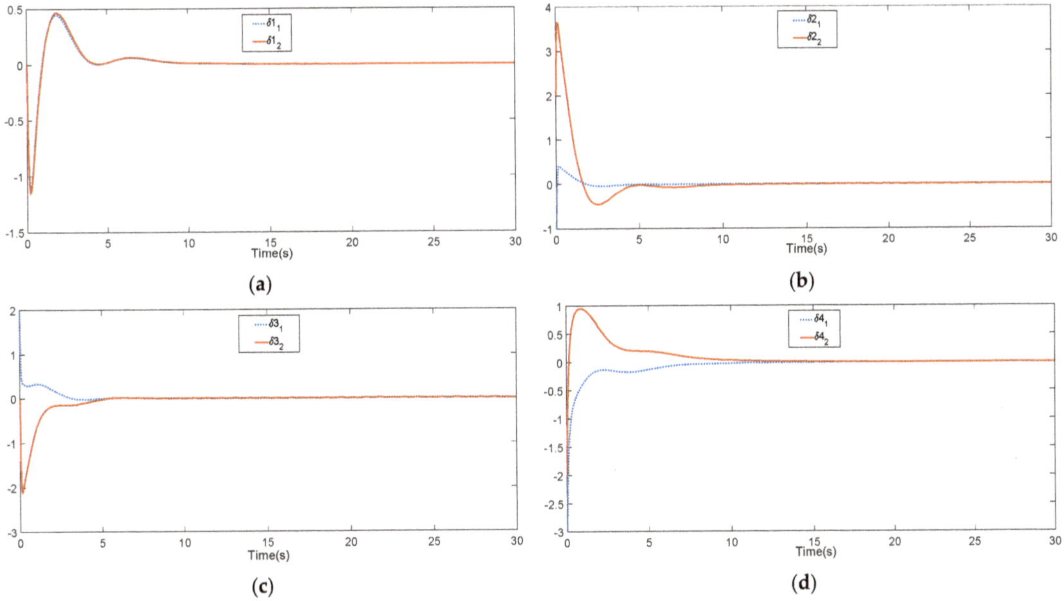

Figure 2. The local neighborhood error vector of each agent: (**a**) The node 1 agent; (**b**) The node 2 agent; (**c**) The node 3 agent; (**d**) The node 4 agent.

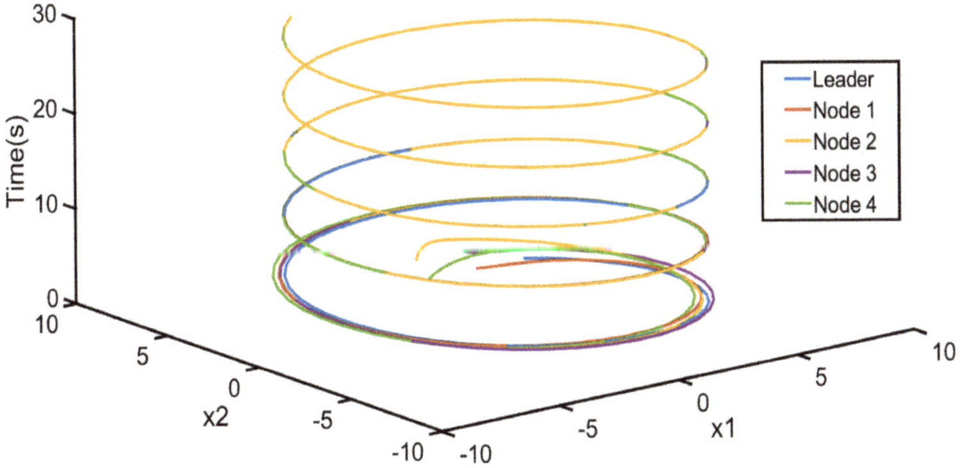

Figure 3. The 3-D phase plane plot of the system's evolution for agents 1, 2, 3, 4 and leader 0.

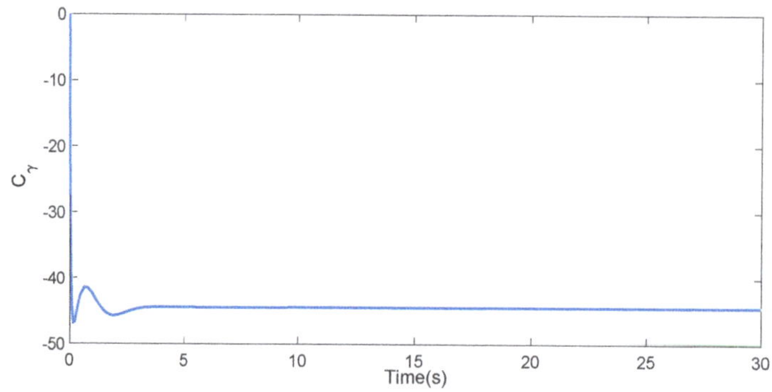

Figure 4. The C_γ of node 1 agent.

6. Conclusions

This paper provides a novel idea for the synchronization control of Multiple Autonomous Unmanned linear systems, in which the local neighborhood error dynamic's coupling part is considered as the virtual external disturbance, so as to decouple the multi-agent cooperative control problem into a relatively independent bounded L_2-gain problem for coupling attenuation. The optimal control theory and differential game theory is utilized to formulate the bounded L_2-gain problem into a centralized multi-player zero-sum game. It is shown that the solution to the multi-player zero-sum game is equivalent to the solution of a coupled HJ equation. It is also shown that the coupled HJ equation can be transformed into an algebraic Riccati equation (ARE) and the solution guarantees the ultimately uniform boundedness (UUB) of the local neighborhood error vector under conservative conditions. The law of parameters selection is summarized. The simulation results show that the proposed method can ensure that the local neighborhood error vectors converge asymptotically to the origin, that is, the multiple autonomous unmanned linear systems can achieve final synchronization, which demonstrates that the UUB of errors is conservative. Meanwhile, the bounded L_2-gain condition for the coupling attenuation can be guaranteed.

This proposed method is suitable for the cooperative consensus control of various homogeneous multiple autonomous unmanned linear systems, such as an underwater robot swarm and aerial UAV swarm. Future work will focus on extending this method to nonlinear multi-agent systems and more serious models with uncertainties.

Author Contributions: Conceptualization, Y.L. and B.W.; methodology, Y.L.; software, Y.L.; validation, Y.L. and Y.C.; formal analysis, Y.C.; investigation, Y.L.; resources, B.W.; data curation, Y.L.; writing—original draft preparation, Y.L.; writing—review and editing, Y.L.; visualization, Y.L. and B.W.; supervision, B.W.; project administration, Y.C.; funding acquisition, B.W. All authors have read and agreed to the published version of the manuscript.

Funding: This research was funded by the National Natural Science Foundation of China (NSFC), grant number 51777058, and in part by the Six Talent Peaks Project in the Jiangsu province, grant number XNY-010.

Institutional Review Board Statement: Not applicable.

Informed Consent Statement: Not applicable.

Data Availability Statement: Not applicable.

Conflicts of Interest: The authors declare no conflict of interest.

References

1. Zhou, Y.; Li, D.; Gao, F. Optimal synchronization control for heterogeneous multi-agent systems: Online adaptive learning solutions. *Asian J. Control* **2021**. [CrossRef]
2. Jing, G.; Zheng, Y.; Wang, L. Consensus of Multiagent Systems With Distance-Dependent Communication Networks. *IEEE Trans. Neural Netw. Learn. Syst.* **2016**, *28*, 2712–2726. [CrossRef] [PubMed]
3. Liu, J.; Dai, M.-Z.; Zhang, C.; Wu, J. Edge-Event-Triggered Synchronization for Multi-Agent Systems with Nonlinear Controller Outputs. *Appl. Sci.* **2020**, *10*, 5250. [CrossRef]
4. Shi, H.; Hou, M.; Wu, Y. Distributed Control for Leader-Following Consensus Problem of Second-Order Multi-Agent Systems and Its Application to Motion Synchronization. *Appl. Sci.* **2019**, *9*, 4208. [CrossRef]
5. Vamvoudakis, K.G.; Lewis, F.L.; Hudas, G.R. Multi-agent differential graphical games: Online adaptive learning solution for synchronization with optimality. *Automatica* **2012**, *48*, 1598–1611. [CrossRef]
6. Vamvoudakis, K.G.; Lewis, F.L. Multi-agent differential graphical games. In Proceedings of the 30th Chinese Control Con-ference, Yantai, China, 22–24 July 2011.
7. Liu, J.; Xu, F.; Lin, S.; Cai, H.; Yan, S. A Multi-Agent-Based Optimization Model for Microgrid Operation Using Dynamic Guiding Chaotic Search Particle Swarm Optimization. *Energies* **2018**, *11*, 3286. [CrossRef]
8. Lowe, R.; Wu, Y.; Tamar, A.; Harb, J.; Abbeel, P.; Mordatch, I. Multi-agent actor–critic for mixed cooperative-competitive environments. *arXiv* **2017**, arXiv:1706.02275. [CrossRef]
9. Liu, Y.; Geng, Z. Finite-time optimal formation control of multi-agent systems on the Lie group SE(3). *Int. J. Control* **2013**, *86*, 1675–1686. [CrossRef]
10. Zhang, W.; Hu, J. Optimal multi-agent coordination under tree formation constraints. *IEEE Trans. Autom. Control.* **2008**, *53*, 692–705. [CrossRef]
11. Lin, W.; Zhao, W.; Liu, H. Robust Optimal Formation Control of Heterogeneous Multi-Agent System via Reinforcement Learning. *IEEE Access* **2020**, *8*, 218424–218432. [CrossRef]
12. Arulkumaran, K.; Deisenroth, M.P.; Brundage, M.; Bharath, A.A. Deep Reinforcement Learning: A Brief Survey. *IEEE Signal Process. Mag.* **2017**, *34*, 26–38. [CrossRef]
13. Mnih, V.; Kavukcuoglu, K.; Silver, D.; Rusu, A.A.; Veness, J.; Bellemare, M.G.; Graves, A.; Riedmiller, M.; Fidjeland, A.K.; Ostrovski, G.; et al. Human-level control through deep reinforcement learning. *Nature* **2015**, *518*, 529–533. [CrossRef] [PubMed]
14. Vinyals, O.; Babuschkin, I.; Czarnecki, W.M.; Mathieu, M.; Dudzik, A.; Chung, J.; Choi, D.H.; Powell, R.; Ewalds, T.; Georgiev, P.; et al. Grandmaster level in StarCraft II using multi-agent reinforcement learning. *Nature* **2019**, *575*, 350–354. [CrossRef]
15. Levine, S.; Pastor, P.; Krizhevsky, A.; Ibarz, J.; Quillen, D. Learning hand-eye coordination for robotic grasping with deep learning and large-scale data collection. *Int. J. Robot. Res.* **2017**, *37*, 421–436. [CrossRef]
16. Tatari, F.; Vamvoudakis, K.G.; Mazouchi, M. Optimal distributed learning for disturbance rejection in networked non-linear games under unknown dynamics. *IET Control Theory Appl.* **2019**, *13*, 2838–2848. [CrossRef]
17. Jiao, Q.; Modares, H.; Xu, S.; Lewis, F.L.; Vamvoudakis, K.G. Multi-agent zero-sum differential graphical games for disturbance rejection in distributed control. *Automatica* **2016**, *69*, 24–34. [CrossRef]
18. Qin, J.; Li, M.; Shi, Y.; Ma, Q.; Zheng, W.X. Optimal Synchronization Control of Multiagent Systems With Input Saturation via Off-Policy Reinforcement Learning. *IEEE Trans. Neural Netw. Learn. Syst.* **2019**, *30*, 85–96. [CrossRef]
19. Kamalapurkar, R.; Walters, P.; Dixon, W. Model-Based Reinforcement Learning for Approximate Optimal Regulation. *Control. Complex Syst.* **2016**, 247–273. [CrossRef]
20. Farrell, J.A. Persistence of excitation conditions in passive learning control. *Automatica* **1997**, *33*, 699–703. [CrossRef]
21. Kamalapurkar, R.; Andrews, L.; Walters, P.; Dixon, W.E. Model-Based Reinforcement Learning for Infinite-Horizon Approximate Optimal Tracking. *IEEE Trans. Neural Networks Learn. Syst.* **2016**, *28*, 753–758. [CrossRef]
22. Yasini, S.; Karimpour, A.; Sistani, M.-B.N.; Modares, H. Online concurrent reinforcement learning algorithm to solve two-player zero-sum games for partially unknown nonlinear continuous-time systems. *Int. J. Adapt. Control Signal Process.* **2014**, *29*, 473–493. [CrossRef]
23. Vamvoudakis, K.G.; Lewis, F.L. Multi-player non-zero-sum games: Online adaptive learning solution of coupled Hamilton–Jacobi equations. *Automatica* **2011**, *47*, 1556–1569. [CrossRef]
24. Bucolo, M.; Buscarino, A.; Fortuna, L.; Frasca, M. LQG control of linear lossless positive-real systems: The continuous-time and discrete-time cases. *Int. J. Dyn. Control* **2022**, *10*, 1075–1083. [CrossRef]
25. Buscarino, A.; Fortuna, L.; Frasca, M.; Rizzo, A. Dynamical network interactions in distributed control of robots. *Chaos: Interdiscip. J. Nonlinear Sci.* **2006**, *16*, 015116. [CrossRef]
26. Khoo, S.; Xie, L.; Man, Z. Robust Finite-Time Consensus Tracking Algorithm for Multirobot Systems. *IEEE/ASME Trans. Mechatron.* **2009**, *14*, 219–228. [CrossRef]
27. Brewer, J. Kronecker products and matrix calculus in system theory. *IEEE Trans. Circuits Syst.* **1978**, *25*, 772–781. [CrossRef]

applied sciences

MDPI

Article

Research on a Measurement Method for the Ocean Wave Field Based on Stereo Vision

Hanyu Sun [1,2,3,*], Guoqing Wu [1,3,*], Xueliang Wang [1,2,3], Tao Zhang [1,2,3], Pu Zhang [1,3], Wei Chen [1,3] and Quanhua Zhu [1,2,3]

1 China Ship Scientific Research Center, Wuxi 214082, China; wangxl@ccsrc.com.cn (X.W.); zhangtao@cssrc.com.cn (T.Z.); zhang_7775@163.com (P.Z.); willenchan@126.com (W.C.); zqh666xx@163.com (Q.Z.)
2 Southern Marine Science and Engineering Guangdong Laboratory (Guangzhou), Guangzhou 511458, China
3 Taihu Laboratory of Deepsea Technological Science, Wuxi 214125, China
* Correspondence: sunhanyu@cssrc.com.cn (H.S.); 15852797071@139.com (G.W.)

Featured Application: This technology can realize near-field wave load monitoring on ships and it can also provide support for short-term wave prediction.

Abstract: The wave parameter is an important environmental input condition. Traditional contact wave measurement methods are unable to meet the requirements of high precision, non-contact, and ship wave field assessment. Alternatively, stereo vision technology can realize a non-contact and mobile form of measurement. However, this technology suffers from poor timeliness and adaptability. This paper proposes a comprehensive wave measurement method that is based on stereo vision, wherein the gridding of siftGPU is used to achieve the fast matching of large images. The whole algorithm can be run within 6 s and it guarantees more than 20,000 feature-matching logarithms. Furthermore, by utilizing the least squares method and sea surface wave surface theory, the sea surface base level can be calculated without control points, along with the inversion of the sea wave parameters (wave height, period, and wave direction) and error point fitting. The rationality and superiority of the algorithm were verified through multiple comparison experiments. Compared with the Richard Brancker Research (RBR) wave height meter, the measurement error of the wave height is less than 10%, the period error is less than 0.5 s, and the wave direction error is less than 10° with the proposed method.

Keywords: stereo vision; gridding siftGPU; wave height; period; wave direction

1. Introduction

Ocean waves indicate a classic and important research direction in the field of physical oceanography [1]. The wave parameter is also an important environmental input condition [2]. The directional spreading of ocean waves plays an important role in various aspects of ocean engineering, such as wave-induced loads, nonlinear wave evolution, and wave breaking [3]. Moreover, consistent sea state time series are essential for building coastal protection or offshore structures [4]. Current wave observation methods can be divided into two categories [5], the first of which is the "point" measurement method that is represented by wave buoys, which requires a fixed mooring line and is troublesome to deploy and maintain [6]. Another one is the "surface" measurement method that is represented by a synthetic aperture radar that obtains wave data depending on the selection of inversion parameters. It has a low measurement accuracy. Owing to such problems in the existing wave observation methods, research on non-contact, mobile and high-precision wave field measurement technology has become one of the important tasks for many countries [7].

Stereoscopic wave measurement technology is a wave measurement technology that uses two photographic cameras to synchronically collect images of the sea's surface. It can directly extract the three-dimensional spatiotemporal distribution of the sea surface fluctuations through the principle of image matching and lens imaging and it obtains the direction spectrum [8]. This technology can effectively make up for the deficiencies of buoys and the remote sensing observation method. Therefore, the development of stereo vision wave measurement technology has important scientific significance and a practical application value for the theoretical development of ocean waves and their application in ocean remote sensing and ocean engineering [9].

The direct linear transformation (DLT) theory was proposed by Abdel-Aziz in 1971; it takes several or even ten days for each pair of images to be completely analyzed, thereby affecting the post-processing time [10]. In 2009, Vies et al. developed a stereo-photographic ocean wave measurement system and performed an experiment on the coast of Scheveningen [11]. They successfully measured the evolution of waves and they found that the measured sea surface area reached 1800 m^2 [12]. Furthermore, in 2011, Wanek developed a fully automated trinocular photography ocean wave observation system (ATSIS) [13]. Different from the other photography systems, ATSIS uses three synchronized cameras to form an image acquisition system in the shape of a "product" and it simulates the measurement scene through the previously recorded camera frame's orientation in the laboratory in order to calibrate the external orientation elements. The ATSIS calibration process is cumbersome and the measurement area is limited by the camera's frame, indeed the ATSIS measurement area is only about 10 m^2 [14]. In 2010, Cai Zhenghan used three synchronous cameras to form an image acquisition system in the shape of a "pin" [15] and then obtained the coordinates of the control point through the means of assessing the underwater makeup. It was necessary to set the calibration points with the current pool and the overall operation was complex [16]. In 2011, Yu Heng of Henan University captured a target image of the Yellow River model water flow vertically through a binocular system. By combining the improved Canny edge detection algorithm with the morphological connected domain segmentation method, he finally extracted and identified the river's potential width and the flow velocity of the Yellow River model [17]. In 2012, Jiang Wenzheng, through his digital photogrammetry wave measurement system, reported a good study on the feature matching of left and right images [18]. Moravec's corner detector is a corner detection operator that is based on gray variance that extracts the points with the maximum and minimum gray variance in four main directions as the feature points [19]. First, the Moravec corner points were extracted from the top layer of a pyramidal image. Then, the left and right images were extracted. Next, the corner points that were extracted from the top-level image of the pyramid were matched based on the principle of a maximum cross-correlation measure, thus obtaining the top-level matching pair set. In the matching pair sets, each matching pair was in the corresponding lower-level area of the pyramid and then matching that was based on a maximum cross-correlation measure was performed. After that, the matching sets of the left and right images were obtained. This feature matching method provides matching pairs as coordinates for the three-dimensional reconstruction of waves and, in this way, realizes sea surface wave measurement. However, the matched results that are obtained via this method have many mismatches. Accordingly, in the future, complex algorithms need to be introduced in order to eliminate the mismatches, alternatively algorithmically accurate wave heights can be introduced in the three-dimensional reconstruction [20]. In 2014, Cui Hao et al. achieved the acquisition of wave crest coordinates based on contour extraction technology, which reduced the influence of the acquisition accuracy on the two-dimensional coordinates of the wave crest in non-contact wave photogrammetry technology, but only the wave's crest was studied [21].

This paper presents a comprehensive process algorithm of wave measurement that is based on stereo vision.

(1) The matching part utilizes the gridding siftGPU method, wherein first the common area of the left and right cameras is determined and, then, grid zoning for the common area is carried out. The grid coordinates of the left camera are used to preliminarily locate the corresponding grid coordinates of the right camera and GPU multithreading processing

is used in order to realize multi feature points and fast matching. Compared with the conventional siftGPU, the matching time with the proposed approach can be controlled below 6 s while the number of matching points can be maintained above 20,000, an outcome that is evidently much better than that of the traditional siftGPU algorithm.

(2) In this paper, the least squares method and wave theory are used to normalize the sea surface-based plane calibration, plane error point fitting, and wave parameter inversion. The design's method is simple and feasible. Compared with the traditional wave height meter, the wave height range is 0.2 m–1 m, the measurement error is less than 10%, and the periodic measurement error is less than 0.5 s.

(3) The Hough transform and least squares method have been used herein to calculate the wave's direction value. It was observed that, for a given wave image, the detection results and the original direction are basically the same.

The rest of this paper is organized as follows. Section 2 introduces the whole process of utilizing the algorithm of wave measurement that is based on stereo vision. Section 3 presents the experimental tests and results. Finally, the conclusions and directions for future work are given in Sections 4 and 5.

2. Materials and Methods

2.1. Stereo Calibration

The purpose of calibration is to obtain a mapping model between the target that is in three-dimensional coordinate space and the two-dimensional image [22]. Essentially, the internal parameters, external parameters, and distortion coefficients that are obtained through this calibration are the basis of the image correction, parameter inversion, and three-dimensional reconstruction. This paper refers to the method that was proposed by Zhang Zhengyou [23] in order to achieve dual-objective targeting. The specific steps are listed as follows:

(1) Construct a checkerboard calibration board with black and white grids, wherein the length and size of the black and white grids are 10 cm and 9 cm × 7 cm, respectively.

(2) Use the binocular camera to take 25 images of the calibration board from different angles.

(3) Use the algorithm to identify the pixel coordinates of the corner points (i.e., the black and white grid intersections) in the image for matching.

(4) Calculate the inversion in order to obtain the internal parameters, external parameters, and lens distortion coefficients of the binocular camera.

2.2. Stereo Matching

2.2.1. Scale-Invariant Feature Transform (SIFT) Feature Extraction and Matching

The first part of the SIFT process is the production of a Gaussian scale space [24], in which each octave is computed by means of the convolution between a down-sampled image and different Gaussian kernels, such as $\mathrm{L}(x, y, \sigma)$ in Equation (1), where $\mathrm{G}(x, y, \sigma)$ is a variable-scale Gaussian kernel and $\mathrm{I}(x, y)$ is an input image.

Next, the second part includes a difference of Gaussian (DoG) scale space [25], this DoG is a close approximation of the Laplacian of Gaussian (LoG), as expressed in Equation (2). Compared with LoG, DoG has lower computational costs and uses approximately the same function to extract stable features.

$$\mathrm{L}(x, y, \sigma) = \mathrm{G}(x, y, \sigma) \times \mathrm{I}(x, y) \quad (1)$$

$$\mathrm{D}(x, y, \sigma) = \mathrm{L}(x, y, k\sigma) - \mathrm{L}(x, y, \sigma) \quad (2)$$

The third part of SIFT contains the pixel and sub-pixel localizations of a keypoint [26]. The extreme pixel is compared with 26 neighboring pixels in a DoG scale space, which is chosen as the candidate pixel keypoint. Then, the localization of the sub-pixel keypoint is implemented using a quadratic approximation, which is computed using the second-order Taylor expansion.

The fourth part of the SIFT algorithm uses 128 dimension descriptors [27] to conduct the feature matching process. The SIFT feature descriptors represent the image gradient magnitudes and orientations in a 4×4 sub-region around the candidate feature.

Unfortunately, this strategy needs high computational power and memory consumption, which limit the use of SIFT in real-time engineering applications [28]. Notably, the above steps of the primal SIFT can run on a GPU in parallel fashion. It is shown that the performance of SiftGPU is nearly real-time for the images and the quality of the features that are extracted by SiftGPU is approximately the same as those that are extracted by the primal SIFT. However, since the GPU's memory is limited, large images need to be down-sampled for size reduction before applying SIFT [29].

Due to the limitations of GPU and SIFT, a gridding siftGPU method is proposed in this work in order to extract and match the features from large images more efficiently.

2.2.2. Gridding SiftGPU Feature Extraction and Matching

First, we used several pairs of wave images to perform the image matching through the use of traditional SIFT, determine the common area of the left and right images, and determine the left image edge points (u_{lmin}, v_{lmin}) and right image edge points (u_{rmax}, v_{rmax}). Here, the common area of the left image is $(u_{lmin} - u_{lmax}, v_{lmin} - v_{lmax})$, while that of right image is $(0 - u_{rmax}, 0 - v_{rmax})$, and the non-overlapping area is cropped.

Subsequently, we divided the common area of the left and right images into matching blocks. At this stage, a specific matching block size was selected according to the actual resolution of the image. It was assumed that the pixel value of a block in the left image was the same as the pixel value in the matching block in the right image. Meanwhile, the size of the areas corresponding to the same matching block are not equal and the pixel coordinates do not have a simple one-to-one correspondence. The calculation method is given in Figure 1:

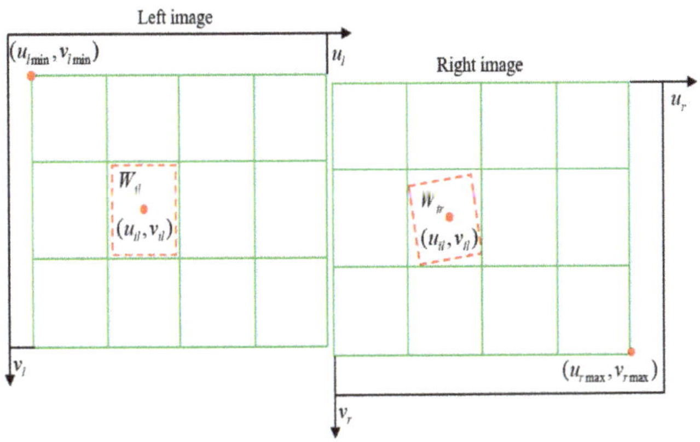

Figure 1. Gridding image and identification of corresponding block.

Considering the camera distortion [30], there are:

$$\begin{cases} x_{il} = x_{0l} + u_{il} + k_1 (x_{0l} + u_{il}) r_{il}^2 \\ y_{il} = y_{0l} + v_{il} + k_1 (y_{0l} + v_{il}) r_{il}^2 \end{cases} \quad (3)$$

In above formula, (u_{il}, v_{il}) are the pixel coordinates of the left image, (x_{0l}, y_{0l}) are the center coordinates of the left image, k is the distortion parameter of the left and right cameras, and r is the distance between the pixel coordinates and center coordinates, $r_{il} = \sqrt{(x+u)^2 + (y+v)^2}$.

As shown in Figure 2, when it is assumed that the vector coordinates of OA are $\vec{r} = (x_{il}, y_{il}, -f)$ and that $\vec{o}$ is the vector coordinates of OO′, then the normal vector of the plane is expressed as follows:

$$\vec{n} = \vec{r} \times \vec{o} \tag{4}$$

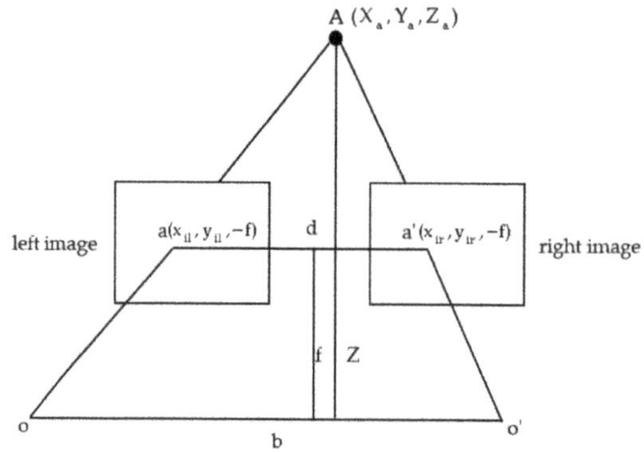

Figure 2. Schematic diagram of the principle of image matching.

Then the normal vector n in the right camera coordinate system is expressed as follows:

$$N_r = R^{-1}N \tag{5}$$

In above formula, R is the rotation matrix of the left camera and N represents the matrix corresponding to normal n. Therefore, the right camera pixel coordinates are:

$$\vec{n}_r(x_{ir}\,\vec{i} + y_{ir}\,\vec{j} - f\vec{k}) = 0 \tag{6}$$

where f is the focus of the camera, N is the vector corresponding to matrix n, and $(x_{ir}, y_{ir}, -f)$ are the coordinates of the right camera.

Considering the camera distortion, there are:

$$\begin{cases} x_{ir} = x_{0r} + u_{ir} + k_1(x_{0r} + u_{ir})r_{ir}^2 \\ y_{ir} = y_{0r} + v_{ir} + k_1(y_{0r} + v_{ir})r_{ir}^2 \end{cases} \tag{7}$$

where k is the distortion parameter of the right camera, r is the distance between the pixel coordinates and the center coordinates, and (x_{0r}, y_{0r}) are the center coordinates of the right image.

In this way, the pixel coordinates of the left and right matching grids were obtained and the position of the left and right corresponding windows was confirmed.

After determining the corresponding matching blocks in the left and right images, we then performed multi-threaded siftGPU accelerated matching on the matching grids in order to achieve fast and accurate matching.

2.3. *Sea Surface Base Plane Calibration*

We then converted the image coordinates to camera coordinates and the specific method for this conversion is described as follows:

Assume that the coordinates of the wave target point in the left camera coordinate system $O_w - X_wY_wZ_w$ are (x_w, y_w, z_w) and that the coordinates of the corresponding image points on the left and right images are (u_l, v_l), (u_r, v_r). According to the conversion relationship, the formula for converting the left and right image points to the left camera coordinate system can be given as:

$$z_{cl}\begin{bmatrix} u_l \\ v_l \\ 1 \end{bmatrix} = H_l \times \begin{bmatrix} x_w \\ y_w \\ z_w \\ 1 \end{bmatrix}, z_{cr}\begin{bmatrix} u_r \\ v_r \\ 1 \end{bmatrix} = H_r \begin{bmatrix} x_w \\ y_w \\ z_w \\ 1 \end{bmatrix} \tag{8}$$

$$H_l = A_l \times \begin{bmatrix} E & 0 \end{bmatrix} \tag{9}$$

$$H_r = A_r \times \begin{bmatrix} R & T \end{bmatrix} \tag{10}$$

In above formula, A_l, A_r are the internal parameter matrices of the left and right cameras, respectively; R, T are the rotation and translation matrices of the dual target, which can be obtained by calibration, and, then, H_l and H_r can be obtained. Expanding the above formula can provide the following formulas:

$$M = \begin{bmatrix} u_l h_{l31} - h_{l11} & u_l h_{l32} - h_{l12} & u_l h_{l33} - h_{l13} \\ v_l h_{l31} - h_{l21} & v_l h_{l32} - h_{l22} & v_l h_{l33} - h_{l23} \\ u_r h_{r31} - h_{r11} & u_r h_{r32} - h_{r12} & u_r h_{r33} - h_{r13} \\ v_r h_{r31} - h_{r21} & v_r h_{r32} - h_{r22} & v_r h_{r33} - h_{r23} \end{bmatrix} \tag{11}$$

$$U = \begin{bmatrix} h_{l14} - u_l h_{l34} \\ h_{l24} - v_l h_{l34} \\ h_{r14} - u_r h_{r34} \\ h_{r24} - v_r h_{r34} \end{bmatrix} \tag{12}$$

where M, U can be obtained from H_l, H_r, and the pixel coordinates (u_l, v_l), (u_r, v_r). Then the camera coordinates of the wave space point (x_w, y_w, z_w) can be obtained through the following formula:

$$\begin{bmatrix} x_w \\ y_w \\ z_w \end{bmatrix} = (M^T \times M)^{-1} M^T U \tag{13}$$

Hereafter, we converted the camera coordinates to the object coordinates; the specific method is as follows:

Select the calm sea images that match. The sea surface base plane equation is given as:

$$ax_w + by_w + z_w + c = 0 \tag{14}$$

In the formula, a and b are the normal vectors in the X and Y directions and c is the average distance from the camera to the sea surface base plane.

Through the transformation of the coordinate system, the coordinates of the feature points in the camera coordinate system are obtained as:

$$B = \begin{bmatrix} x_{w1} & y_{w1} & 1 \\ x_{w2} & y_{w2} & 1 \\ \cdots & \cdots & \cdots \\ x_{wn} & y_{wn} & 1 \end{bmatrix} \tag{15}$$

$$L = \begin{bmatrix} z_{w1} & z_{w2} & \cdots & z_{wn} \end{bmatrix}^T \tag{16}$$

The parameters of the sea surface base plane can be calculated using the least squares method as:

$$[a, b, c] = (B^T \times B)^{-1} \times B^T \times L \tag{17}$$

Normalizing the plane normal vector that was obtained from each image results in:

$$\vec{n} = \frac{ai + bj - k}{\sqrt{a^2 + b^2 + 1^2}} \tag{18}$$

In the formula, $\vec{n}$ is the normal vector of the sea surface base plane.

The spherical coordinate transformation is performed on the normal vector, which is convenient for calculating the changing relationship between the image coordinate system and the object coordinate system.

$$\varphi = \arccos(c/\sqrt{a^2 + b^2 + 1^2}) \tag{19}$$

$$\theta = \arctan(b/a) \tag{20}$$

In these formulas, φ and θ correspond to the spherical coordinates of the sea surface base plane.

Then, the conversion of the rotation and translation matrices from the camera coordinate system to that object coordinate system is expressed as:

$$R_W = \begin{bmatrix} \sin\varphi & -\cos\varphi & 0 \\ \cos\theta\cos\varphi & \cos\theta\sin\varphi & -\sin\theta \\ \sin\theta\cos\varphi & \sin\theta\sin\varphi & \cos\theta \end{bmatrix} \tag{21}$$

$$T_W = R_W\begin{pmatrix}0 & 0 & -c\end{pmatrix}^T \tag{22}$$

Subsequently, the camera coordinates are obtained based on the sea surface base plane equation for the conversion from the camera coordinate system (x_W, y_W, z_W) to the object coordinate system (x_c, y_c, z_c). This conversion equation is:

$$(x_c, y_c, z_c)^T = R_W(x_W, y_W, z_W)^T + T_W \tag{23}$$

2.4. Wave Parameter Inversion

1. Wave Height and Period

The center point of the matching block was selected as the wave inversion point. In order to ensure the accuracy of the calculation, the center point of the four matching blocks was selected in order to calculate the wave's height and the changes in the wave data throughout 10 min was recorded. Next, the spectrum was obtained by the use of the Fourier transform and the main frequency range was selected. Following this, we selected the frequency of 0–0.5 Hz, filtered out the high-frequency part, and performed an inverse Fourier transform in order to obtain the time-domain curve corresponding to the main frequency. Then, we used the zero-crossing method to calculate the wave height and period.

2. Wave Direction

The wave direction calculation method that was employed in this paper is as follows:

Essentially, the Hough transform is a feature extraction method. In this work, the Hough transform was performed on the left image and the specific steps are as follows:

1. Use a two-dimensional Gaussian function to the detect the extreme points. Obtain the coordinates of those extreme points (u_i, v_i).
2. Convert the planar coordinates (u_i, v_i) to polar coordinates (ρ, θ) by using:

$$\rho = u\cos\theta + v\sin\theta \tag{24}$$

3. Calculate the maximum value with a two-dimensional accumulator $A(a, b)$ as:

$$A(a, b) = A(a, b) + 1 \tag{25}$$

The camera resolution is $M_1 \times M_2$ and, therefore, the parameter coordinate range is:

$$\rho \in \left[-2\sqrt{M_1}/2, 2\sqrt{M_2}/2\right] \tag{26}$$

$$\theta \in [-\pi/2, \pi/2] \tag{27}$$

The maximum value corresponding to the accumulator in the statistical parameter coordinate system is $[\rho, \theta]$. Record all of the pixel coordinates on the line.

Next, calculate the corresponding (u_{ih}, v_{ih}) and $(x_{ihc}, y_{ihc}, z_{ihc})$. If $z_{ihc} \geq 0.95 \times z_{max}$, calculate the normal vector of the straight line $\rho = x\cos\theta + y\sin\theta$. On the other hand, if $z_{ihc} \leq 0.95 \times z_{max}$, select the corresponding coordinates to carry out the least squares fitting, so as to calculate the normal vector of the main peak line;

The normal vector of each straight line corresponds to two angles; thus it is necessary to determine the direction of the wave motion. In this paper, true north and counterclockwise rotation are considered as the positive direction when judging the change in the intercept of the wave crest line that was calculated in the frame images afterwards, so as to determine the wave movement direction.

3. Results

3.1. Laboratory Verification

The relevant experiment was carried out in the China Shipbuilding Research Center. The experiment used two cameras with a resolution of 6280 × 3158 in order to obtain the impact pair (shown in the Figure 3). The image shooting time was 17 January 2022. Moreover, a calibration box was used to check the system algorithm in order to measure its accuracy and the specific parameters are as follows:

1. Calibration Results

(a) (b)

Figure 3. Illustration: (**a**) calibration box verification picture 1; (**b**) calibration box verification picture 2.

The calibration results of the internal and external parameters of the camera are in Table 1:

2. Gridding siftGPU Parameter Calculation

For the left picture, $u_{lmin} = 580$ and $v_{lmin} = 158$; whereas, for the right picture, $u_{rmax} = 5700$ and $v_{rmax} = 3000$. Meanwhile, the gridding matching block size was 300 × 300 and the number of matching blocks was 190. Here, we chose the left picture (580, 158) and the right image (0, 0) for verification and the matching results are provided in the following Figure 3.

Table 1. System calibration parameters.

Code No.	Title 2
Left camera internal reference (pixel)	$\begin{bmatrix} 13{,}868.3 & 0 & 2973.9 \\ 0 & 13{,}860.9 & 1614.1 \\ 0 & 0 & 1 \end{bmatrix}$
Right camera internal reference (pixel)	$\begin{bmatrix} 13{,}911.8 & 0 & 3047.1 \\ 0 & 13{,}903.6 & 1691.9 \\ 0 & 0 & 1 \end{bmatrix}$
Distortion coefficient of left camera (pixel)	$\begin{bmatrix} -0.11 & -0.58 \end{bmatrix}$
Distortion coefficient of right camera (pixel)	$\begin{bmatrix} -0.15 & 1.18 \end{bmatrix}$
Common area boundary parameters (pixel)	$\begin{bmatrix} 580 & 158 \\ 5700 & 3000 \end{bmatrix}$
Rotation matrix (pixel)	$\begin{bmatrix} 0.9997 & -0.0254 & -0.0009 \\ 0.0254 & 0.9996 & 0.0110 \\ -0.0011 & -0.0110 & 0.9999 \end{bmatrix}$
Translation matrix (pixel)	$\begin{bmatrix} -267.56134 \\ -1.75769 \\ 15.63543 \end{bmatrix}$

Following this, siftGPU and gridding siftGPU were used to match image 1 and image 2 and the matching result is shown in the following Figures 4–7.

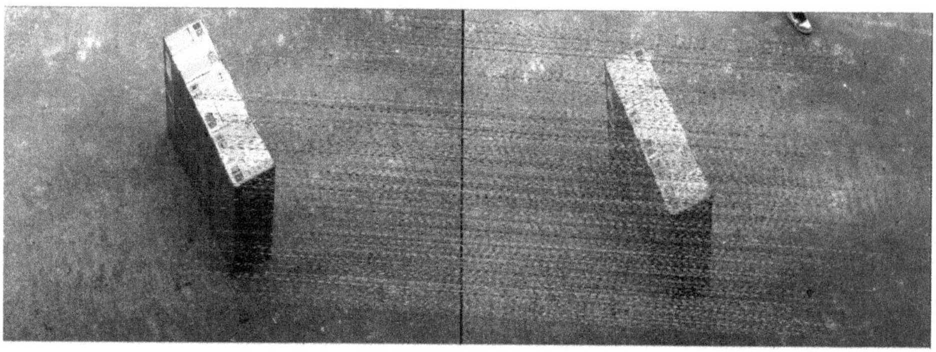

Figure 4. Image 1 matching numbers based on siftGPU.

Figure 5. Image 1 matching numbers based on gridding siftGPU.

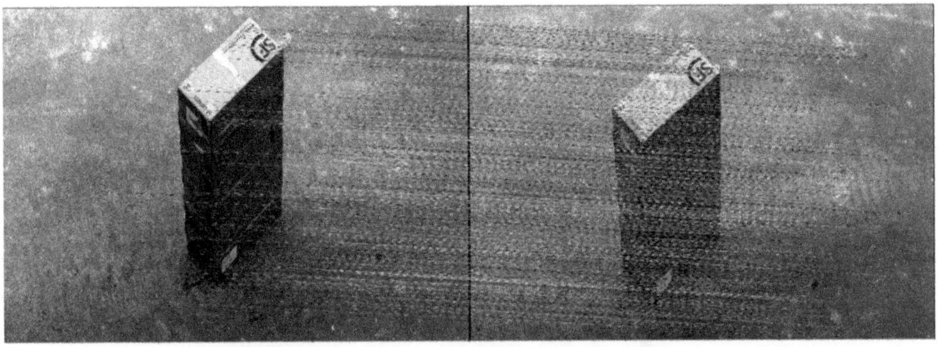

Figure 6. Image 2 matching numbers based on siftGPU.

Figure 7. Image 2 matching numbers based on gridding siftGPU.

It can be seen from Figures 4–7 that the matching number of feature points that are based on the gridding siftGPU algorithm is obviously better than that of the points that are based on the siftGPU algorithm. The calculation results are shown in the Table 2.

Table 2. Matching time and feature points numbers.

Code No.	siftGPU	Gridding siftGPU
Threshold	0.5	0.5
Image 1 matching time (s)	31.7	6
Image 1 matching no.	3003	24,424
Image 2 matching time (s)	29.9	5.8
Image 2 matching no.	2996	24,321

By comparison, the matching time of the gridding siftGPU was 1/6 of that of the siftGPU, whereas the match quantity that was based on the gridding siftGPU was 8 times that which was based on siftGPU.

3. Calibration Results for the Base Plane

Twenty feature points were selected in image 2 in order to calculate the base plane. The selected area is the ground feature in image 2, which is shown in the Figure 8:

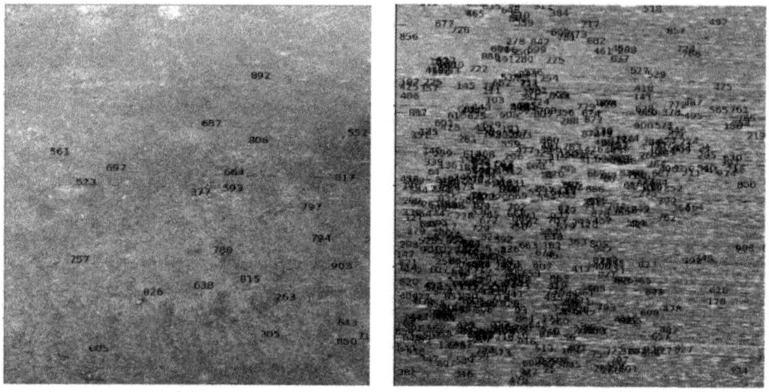

(a) (b)

Figure 8. Illustration: (**a**) sea surface base plane calibration image 1; (**b**) sea surface base plane calibration image 2.

The code numbers and coordinates of the selected feature points are shown in Table 3.

Table 3. Sea surface base plane calibration feature point coordinates.

Code No.	ul (mm)	vl (mm)	ur (mm)	vr (mm)
561	462.869	1400.559	117.233	1557.912
523	602.897	1604.144	244.895	1763.765
757	569.487	2124.842	181.769	2283.181
697	758.91	1509.639	404.889	1673.864
605	676.614	2716.894	254.76	2878.093
826	956.753	2342.16	553.65	2510.346
687	1266.115	1209.11	926.654	1386.103
377	1207.998	1668.579	843.952	1843.557
638	1228.33	2294.944	828.212	2470.034
892	1526.094	886.002	3618.396	2288.871
664	1388.562	1534.327	1030.845	1714.759
593	1378.192	1638.199	1014.254	1818.597
780	1326.869	2057.044	940.375	2235.672
806	1513.996	1317.925	1168.24	1501.776
815	1471.925	2253.082	1072.941	2435.356
867	4628.848	346.125	4336.211	608.107
872	4956.792	1102.291	4623.232	1372.855
553	4666.975	1550.444	4307.868	1814.572
339	4113.929	1368.649	3762.998	1618.408
927	4781.349	1701.514	4414.901	1968.805

The calculation parameters of the sea surface base plane are as follows:

$$a = 2.3715,\ b = -1.5122,\ c = -8428.2437$$

The calculation parameters of R_w and T_w are:

$$R_w = \begin{bmatrix} -0.9983 & -0.059 & 0 \\ -0.042 & 0.717 & -0.696 \\ 0.041 & -0.695 & 0.718 \end{bmatrix}$$

$$T_w = \begin{bmatrix} 0 \\ 5868.045 \\ 6049.904 \end{bmatrix}$$

Additionally, the scatter diagram is shown in Figure 8.

As shown in Figure 9, the original 3D point images of image 1 and image 2 are consistent with the original image. The blue points represent the ground features and the yellow features represent the calibration box features. The white area is the unrecognized area, due to the strong light.

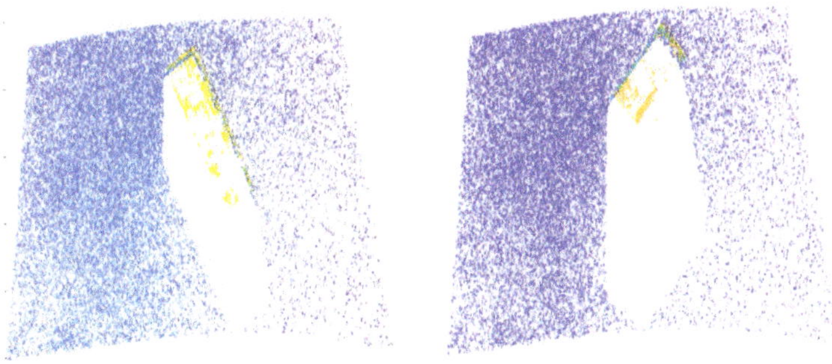

Figure 9. Illustration: (**a**) original 3D point images 1; (**b**) original 3D point images 2.

Next, 14 points were selected in order to verify the measurement accuracy of the system. The obtained measurement results were compared with the tape (the minimum scale was 1 mm) and the error was basically controlled within 1%. The results are shown in Tables 4 and 5.

Table 4. Measurement accuracy verification of the calibration box image 1.

Code No.	x (mm)	y (mm)	z (mm)	Calibration Box Width (mm)	Error
7	63.67009	483.8519	690.5644	690	0.08%
8	102.5964	349.7061	692.4865	690	0.36%
9	157.9489	−134.73	689.9507	690	0.01%
10	−127.546	750.4248	696.5345	690	0.95%
11	−127.597	745.3457	691.4332	690	0.21%
12	56.47901	491.2691	685.9605	690	0.59%
13	102.3608	353.218	695.7244	690	0.83%
14	−133.579	377.7311	691.3577	690	0.20%
15	92.31968	365.5422	690.5268	690	0.08%
16	232.3613	−295.85	688.9657	690	0.15%
17	114.8581	−181.829	690.5955	690	0.09%
20	42.98887	480.5747	692.0436	690	0.30%
21	−109.301	716.635	694.6232	690	0.67%
22	119.2851	316.8094	689.2604	690	0.11%

Table 5. Measurement accuracy verification of calibration box image 2.

Code No.	x (mm)	y (mm)	z (mm)	Calibration Box Height (mm)	Error
1	−164.961	424.689	1202.88	1204	0.09%
2	−101.668	237.8386	1202.787	1204	0.10%
3	−206.176	433.3536	1208.533	1204	0.38%
4	−155.174	375.2888	1202.448	1204	0.13%
5	−102.034	231.6729	1197.777	1204	0.52%
6	−250.352	385.6709	1207.963	1204	0.33%
7	−155.551	376.6541	1203.015	1204	0.08%
8	−133.514	398.3701	1200.575	1204	0.28%

Table 5. *Cont.*

Code No.	x (mm)	y (mm)	z (mm)	Calibration Box Height (mm)	Error
9	−183.244	358.0353	1205.239	1204	0.10%
10	−229.913	410.1861	1208.479	1204	0.37%
11	221.4955	801.7397	1185.754	1204	1.52%
12	−194.046	434.6676	1207.884	1204	0.32%
13	−233.126	391.6223	1211.303	1204	0.61%
14	−171.595	490.7635	1206.244	1204	0.19%

3.2. Offshore Test Verification

The sea test was performed on the Qingdao Blue Valley No. 1 fixed platform on 19 January 2022. The image was captured simultaneously by two QHY26m industrial cameras and the lens that was used was a Nikon 50 mm/1.4D fixed-focus. The pixels of the QHY26M industrial camera were arranged to a value of 6280 × 3158, the pixel size was 3.76 um, and the height of the camera from the sea surface was about 10 m (relative to the tide level). The exposure time was set to 2 ms, the sampling frequency was 2 Hz, and the test position was the hatch of the platform. The test interval was between 14:00–16:00 and, in each observation, the data were collected for 10 min. The results of this work are compared with an RBR wave height meter's data for verification. Furthermore, the test parameters are consistent with the laboratory verification parameters and the test pictures are shown in Figure 10:

(a) (b)

Figure 10. Illustration: (**a**) stereo vision ocean wave observation system; (**b**) RBR wave height meter.

The size of the field was $X_{min} = -1500$ mm, $X_{max} = 1000$ mm, $Y_{min} = -1000$ mm, and $Y_{max} = 1000$ mm. The coordinates of the center point (0, 0) were selected in order to verify the accuracy of the measured wave height and period. The measurement results are shown in the Figure 11:

As shown in the Figure 11, in the same time interval, the time–domain variation curve of wave height as measured at grid point (0, 0) with the proposed stereo vision wave measuring system is consistent with the time–domain variation trend of the wave height as measured by an RBR wave height meter. Moreover, the measurement results at grid point (−320, 0) are consistent with those at grid point (0, 0), in the same time interval.

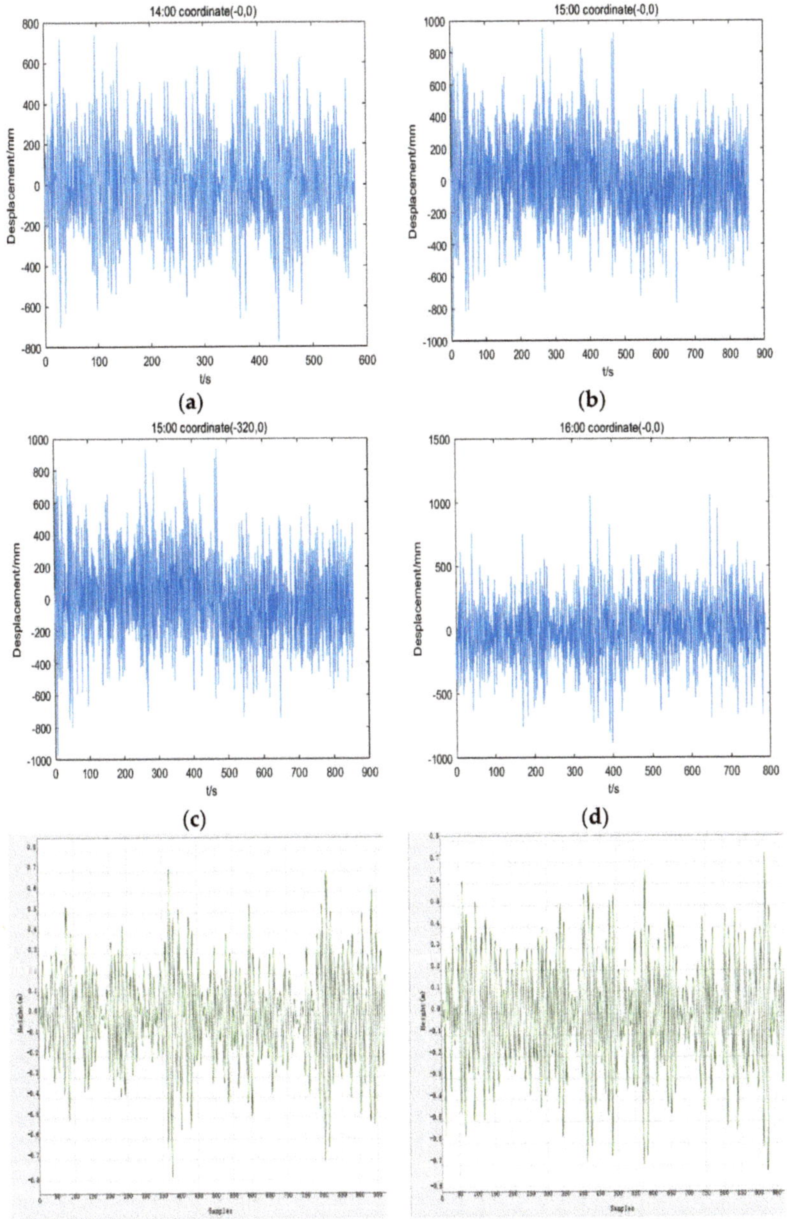

Figure 11. *Cont.*

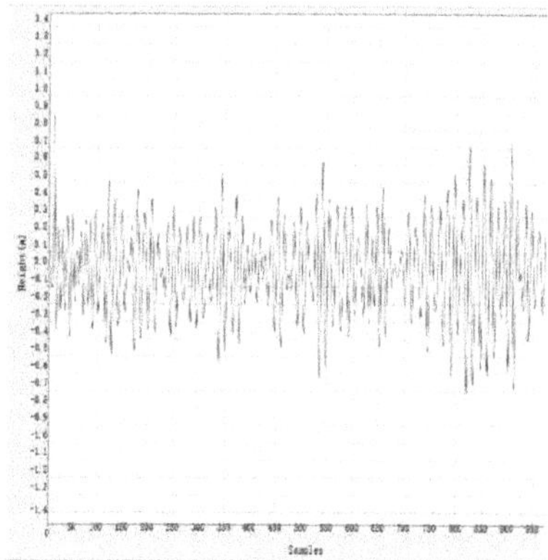

(g)

Figure 11. Illustration: (**a**) at 14:00 with stereo vision, variation in coordinate (0, 0) of sea surface displacement with time; (**b**) at 15:00 with stereo vision, variation in coordinate (0, 0) of sea surface displacement with time; (**c**) at 16:00 with stereo vision, variation in coordinate (−320, 0) of sea surface displacement with time; (**d**) at 16:00 with stereo vision, variation in coordinate (0, 0) of sea surface displacement with time; (**e**) at 14:00 with RBR wave high meter of sea surface displacement with time; (**f**) at 15:00 with RBR wave high meter of sea surface displacement with time; (**g**) at 16:00 with RBR wave high meter of sea surface displacement with time.

The calculation results are shown in Table 6:

Table 6. Measurement accuracy verification.

Time	Parameter	RBR	Stereo Vision	Error
14:00	1/3 wave height (m)	0.870	0.857	1.5%
	1/10 wave height (m)	1.160	1.107	4.6%
	1/100 wave height (m)	1.498	1.447	3.4%
	Average wave height (m)	0.556	0.552	0.7%
	Wave period (s)	6.167	6.615	0.45
15:00	1/3 wave height (m)	0.945	0.942	0.3%
	1/10 wave height (m)	1.203	1.220	1.4%
	1/100 wave height (m)	1.476	1.594	8.0%
	Average wave height (m)	0.615	0.622	0.1%
	Wave period (s)	6.903	7.054	0.35
16:00	1/3 wave height (m)	1.033	1.029	0.3%
	1/10 wave height (m)	1.309	1.300	1.4%
	1/100 wave height (m)	1.704	1.639	8.0%
	Average wave height (m)	0.689	0.716	0.1%
	Wave period (s)	6.407	6.728	0.32

As shown in the Table 7, the wave height error that was based on binocular stereo vision is less than 10% and the period is less than 0.5 s, relative to the results of the RBR wave height meter.

Table 7. Data comparison for different measured coordinates.

Time	Parameter	Coordinate (−320, 0)	Coordinate (0, 0)	Error
14:00	1/3 wave height (m)	0.942	0.939	0.3%
	1/10 wave height (m)	1.220	1.218	1.4%
	1/100 wave height (m)	1.594	1.602	8.0%
	Average wave height (m)	0.622	0.625	0.1%
	Wave period (s)	7.054	7.253	0.35

The wave direction diagram is shown in the Figure 12:

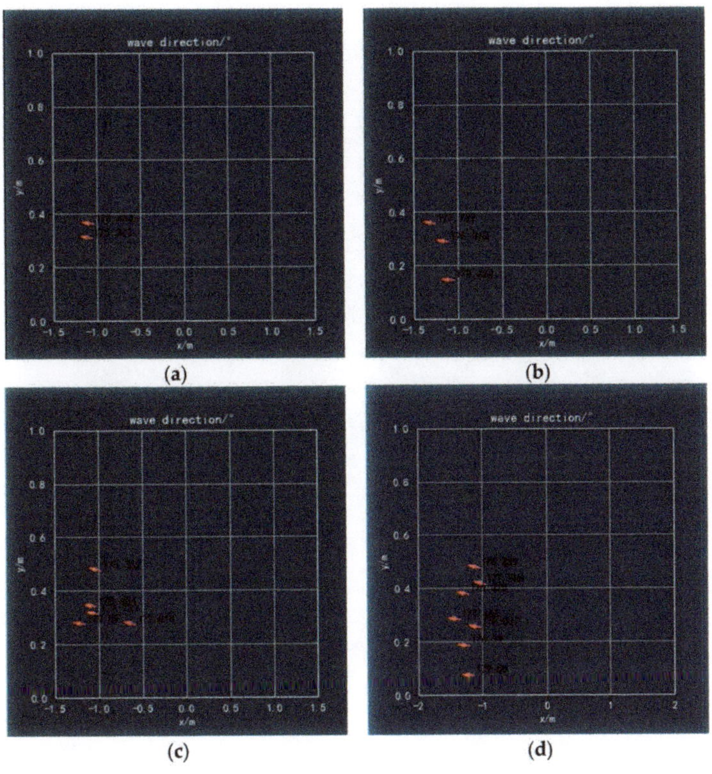

Figure 12. Illustration: (**a**) wave direction diagram of the first image frame; (**b**) wave direction diagram of the second image frame; (**c**) wave direction diagram of the third image frame; (**d**) wave direction diagram of the thirteenth image frame.

As shown in Figure 12, the wave direction of the normal vector was basically consistent with the actual movement direction of the wave. The wave direction measurement error is ±10 degrees and the resolution is 1 degree. These findings imply that the measurement results are accurate. The calculation results are shown in Table 8:

Table 8. Measurement accuracy verification of wave direction.

Time	Parameter	RBR	Stereo Vision	Error
14:00	Wave direction (degrees)	170	175	5
15:00	Wave direction (degrees)	171	177	6
16:00	Wave direction (degrees)	170	172	2

A 3D reconstruction of the wave field is shown in Figure 13:

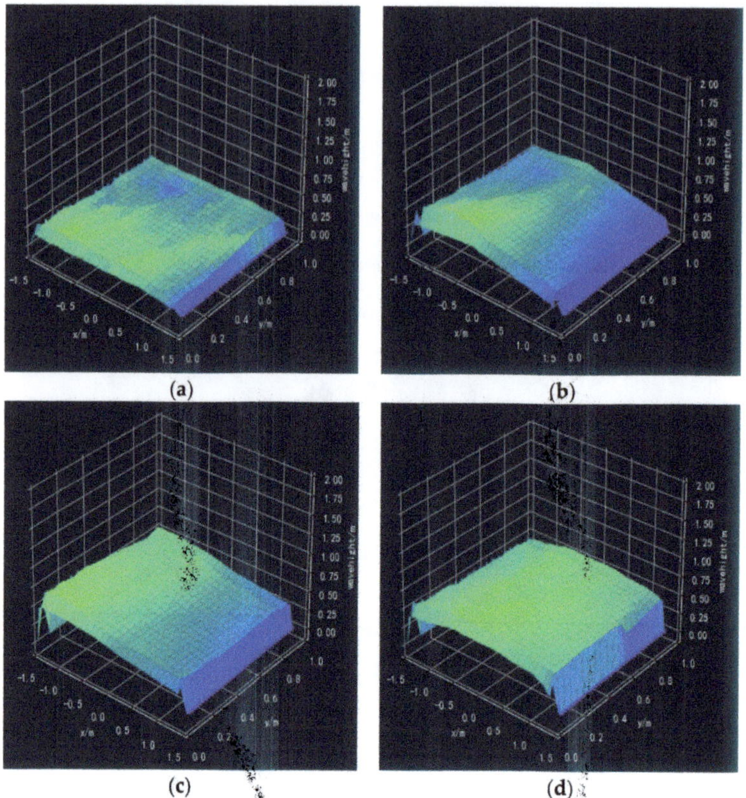

Figure 13. Illustration: (**a**) wave field 3D reconstruction of the first frame image; (**b**) wave field 3D reconstruction of the second frame image; (**c**) wave field 3D reconstruction of the third frame image; (**d**) wave field 3D reconstruction of the fourth frame image.

As is evident from Figure 12, the proposed method can effectively realize the three-dimensional reconstruction of an ocean wave field.

The wave height value of the 3D inversion is basically the same as the measurement result of the RBR wave height meter.

4. Discussion

Based on the results that were achieved by utilizing stereo vision in wave-height measurement, the following conclusions have been drawn:

(1) Relying on the actual sea trial measurement results, it can be confirmed that the stereo vision wave measurement method can enable the accurate measurement of wave parameters (wave height, period, and wave direction) [30]. This is because the stereo vision technology uses the parallax formula to calculate the distance to the measured object, while the height of the object can be measured through corresponding coordinate conversion. Through the calculation of the elevation value of a fixed area, the wave height and the period of the sea wave can be analyzed. Furthermore, the wave direction can be detected using the normal vector of the wave crest line in the image, therefore it is necessary to recognize and fit the line features.

(2) In order to realize a real-time operation of the whole process algorithm, the richness of the wave field's features, and the accuracy of the measurement results, it is necessary

to use a high-resolution camera for the measurement. However, this results in a too long matching time and very high computer memory costs. In this regard, the gridding siftGPU algorithm that is proposed in this paper considers that the common area of the left and right images can be gridded without down-sampling and the number of feature pairs can be matched using GPU multithreading processing, hence the matching is faster.

At present, radar is used for onboard wave measurements, however radar measurement is expensive and it has a low level of accuracy [31]. The fixed-point measurement methods mainly include ADCP and the wave heigh meter, both of which have high requirements for the water's depth and they are also troublesome to place and recover. ADCP needs to be installed on the ground, while the wave heigh meter requires mooring line fixation [32]. In addition, both of these methods can only realize single point measurement and the measurement results will degrade due to the process of wave breaking under strong wave conditions. In contrast, stereo vision wave measurement technology can realize a non-contact and high-precision measurement. The main limitation of the approach that is presented in this paper is the high requirements of computer memory and GPU. Besides this, the internal and external parameters of the camera need to be calibrated manually before the measurement is undertaken. The calibration parameters also need to be kept unchanged throughout the actual application.

Binocular stereo wave measurement technology is a promising field, nevertheless there are still some problems that need to be solved in order to fully enable this technology for future engineering application [33]. These include the system's motion compensation and applicability to complex lighting conditions (rainy and foggy days). The pre-calibration part can form a standard library for the internal and external parameters of the camera, which can be called directly in the practical application, so as to adapt to different measurement distances. Under complex lighting conditions, the illumination adaptability of the algorithm needs to be further studied. The relationship between the illumination intensity and the best aperture and focal lengths can be trained by deep learning in order to realize all-weather automatic focusing [34]. The system needs to consider the motion compensation algorithm and hardware for the motion of the platform. Taking a ship as an example, rolling and surge will not affect the measurement accuracy of the system and the heave direction needs to be compensated for by the integration of the results of the acceleration measurement, such as rolling and pitch. For yaw, a three degree of freedom motion compensation platform needs to be developed for the hardware's compensation. Furthermore, with the development of ship and marine intelligent chemical engineering, marine measurement technology that is based on vision will become more and more important.

5. Conclusions

This paper presents a comprehensive processing algorithm for wave measurement that is based on stereo vision.

(1) Compared with the conventional siftGPU, based on 6280 × 4210 image processing, the matching time can be controlled below 6 s and the number of matching points can be kept above 20,000, indicating much better performance than that of the traditional siftGPU algorithm.

(2) In this paper, the least squares method and wave theory have been used to normalize the sea surface base plane calibration, plane error point fitting and wave parameter inversion. This design method is simple and feasible. Compared with RBR wave height measurement, the wave height range is 0.2 m–1 m, the measurement error is less than 10%, and the periodic measurement error is less than 0.5 s.

(3) In this paper, the Hough transform and least squares method were used to calculate the wave direction value. Because the RBR wave height instrument does not have the function of wave direction measurement, compared with the original wave image, the detection results and original direction are basically the same.

Author Contributions: Conceptualization, H.S.; methodology, H.S.; software, W.C.; validation, Q.Z.; formal analysis, X.W.; investigation, G.W.; resources, P.Z.; data curation, T.Z.; writing—original draft preparation, H.S.; writing—review and editing, H.S.; visualization, G.W.; supervision, T.Z.; project administration, X.W. All authors have read and agreed to the published version of the manuscript.

Funding: This research was funded by the Key Special Project for Introduced Talents Team of Southern Marine Science and Engineering Guangdong Laboratory (Guangzhou), grant number GML2019ZD-0502, and the High Tech Ship Scientific Research Project of the Ministry of Industry and Information Technology, grant number [2019]357.

Institutional Review Board Statement: Not applicable.

Informed Consent Statement: Not applicable.

Data Availability Statement: The data that were used to support the findings of this study are included within the article.

Acknowledgments: The authors would like to thank the authors of SiftGPU and SIFT for making their algorithms available as open-sources, this was really helpful to the research that is described in this paper.

Conflicts of Interest: The authors declare no conflict of interest.

References

1. Lin, Z.X.; Adcock, A.A.T.; McAllister Mark, L. Estimating ocean wave directional spreading using wave following buoys: A comparison of experimental buoy and gauge data. *J. Ocean Eng. Mar. Energy* **2022**, *8*, 93–97. [CrossRef]
2. Christoph, J.; Berkenbrink, C.; Stumpe, B. Prediction and reconstruction of ocean wave heights based on bathymetric data using LSTM neural networks. *Ocean Eng.* **2021**, *232*, 109046.
3. Makar, A. Determination of the Minimum Safe Distance between a USV and a Hydro-Engineering Structure in a Restricted Water Region Sounding. *Energies* **2022**, *15*, 2441. [CrossRef]
4. Specht, C.; Lewicka, O.; Specht, M.; Dąbrowski, P.; Burdziakowski, P. Methodology for Carrying out Measurements of the Tombolo Geomorphic Landform Using Unmanned Aerial and Surface Vehicles near Sopot Pier, Poland. *J. Mar. Sci. Eng.* **2020**, *8*, 384. [CrossRef]
5. Ying, W. Research on three-dimensional observation technology and application of shore based digital photography. *Ocean. Univ. China* **2015**, *58*, 239–250.
6. Rossi, G.B.; Cannata, A.; Iengo, A.; Migliaccio, M.; Nardone, G.; Piscopo, V.; Zambianchi, E. Measurement of Sea Waves. *Sensors* **2022**, *22*, 78. [CrossRef]
7. Wang, H.; Mouche, A.; Husson, R.; Grouazel, A.; Chapron, B.; Yang, J. Assessment of Ocean Swell Height Observations from Sentinel-1A/B Wave Mode against Buoy In Situ and Modeling Hindcasts. *Remote Sens.* **2022**, *14*, 862. [CrossRef]
8. Bergamasco, F.; Benetazzo, A. Spatially Distributed Sea Wave Measurements. *J. Mar. Sci. Eng.* **2021**, *9*, 238. [CrossRef]
9. Sun, X.; Jiang, Y.; Ji, Y.; Fu, W.; Yan, S.; Chen, Q.; Yu, B.; Gan, X. *Distance Measurement System Based on Binocular Stereo Vision In IOP Conference Series: Earth and Environmental Science*; IOP Publishing: Bristol, UK, 2019; Volume 252, pp. 1500–1504.
10. Xie, H.Z.; Yao, H.X.; Zhou, S.C.; Zhang, S.P.; Tong, X.J.; Sun, W.X. Toward 3D object reconstruction from stereo images. *Neurocomputing* **2021**, *463*, 950–955. [CrossRef]
11. Nagano, A. Three-dimensional videography using omnidirectional cameras: An approach inspired by the direct linear transformation method. *J. Biomech.* **2021**, *128*, 110722. [CrossRef]
12. De Vries, S.; Hill, D.F.; De Schipper, M.A.; Stive, M.J.F. Remote sensing of surf zone waves using stereo imaging. *Coast. Eng.* **2011**, *58*, 239–250. [CrossRef]
13. De Vries, S.; Hill, D.; De Schipper, M.A.; Stive, M.J.F. Using Stereo Photogrammetry to Measure Coastal Wavs. *J. Coast. Res.* **2009**, *58*, 1484–1488.
14. Justin, M.W.; Chin, H.W. Automated trinocular stereo imaging system for three-dimensional surface wave measurements. *Ocean. Eng.* **2005**, *33*, 255–259.
15. Guillaume, G.; Ludovic, C.; Damien, C.; Laurent, D. Free surface measurement by stereo-refraction. *Exp. Fluids Exp. Methods Appl. Fluid Flow* **2013**, *54*, 1900–1904.
16. Cai, Z.H.; Dong, D.J.; Zhou, Y.Y. Research and development of stereo sea state photography technology. *Taiwan Cent. Meteorol. Adm.* **2010**, *36*, 101–106.
17. Yu, H. Research and Application of Image Measurement Method for River Regime Width and Surface Velocity of Yellow River Model. Master's Thesis, Henan University, Henan, China, 2011.
18. Jiang, W.Z. Digital stereo photography wave measurement technology. Ph.D. Thesis, Ocean University of China, Qingdao, China, 2012.
19. Zheng, K.; Jiang, W.Z.; Lu, X.; Wang, S.L. Three dimensional reconstruction of wave surface based on binocular stereo vision. *Sci. Technol. Eng.* **2021**, *21*, 735–740.
20. Wei, S. Research on Wave Contour Measurement Based on Binocular Vision. Master's Thesis, Harbin Engineering University, Harbin, China, 2015.

21. Li, T.; Zhou, M. ECG classification using wavelet packet entropy and random forests. *Entropy* **2016**, *18*, 285. [CrossRef]
22. Hagemann, A.; Knorr, M.; Janssen, H.; Stiller, C. Inferring Bias and Uncertainty in Camera Calibration. *Int. J. Comput. Vis.* **2022**, *130*, 17–32. [CrossRef]
23. Mei, Z.; Wang, Y.; Li, Q.Y. Research on Moving Target Detection and Tracking Technology in Sports Video Based on SIFT Algorithm. *Adv. Multimed.* **2022**, *2022*, 1900–1905. [CrossRef]
24. Zhang, Y.M.; Shu, J.; Hu, L.; Zhou, Q.; Du, Z.R. A Ship Target Tracking Algorithm Based on Deep Learning and Multiple Features. In *Twelfth International Conference on Machine Vision (ICMV 2019)*; SPIE: Washington, WA, USA, 2020; Volume 11433, pp. 19–26.
25. Wang, Y.C.; Yuan, Y.J.; Zhao, L. Fast SIFT Feature Matching Algorithm Based on Geometric Transformation. *IEEE Access* **2020**, *8*, 735–740. [CrossRef]
26. Sun, H.Q.; Li, B.G.; Jin, Y.Y. Research on Fingerprint Cipher Based on SIFT Feature Matching Principle. *Int. J. High. Educ. Teach. Theory* **2021**, *2*, 16–20.
27. Yang, J.R.; Huang, J.W.; Jiang, Z.Y.; Dong, S.B.; Tang, L.Q.; Liu, Y.P.; Liu, Z.J.; Zhou, L.C. 3D SIFT aided path independent digital volume correlation and its GPU acceleration. *Opt. Lasers Eng.* **2021**, *136*, 1177–1180. [CrossRef]
28. Li, J.S.; Yun, P. GPU-based parallel optimization for real-time scale-invariant feature transform in binocular visual registration. *Pers. Ubiquitous Comput.* **2019**, *23*, 465–474. [CrossRef]
29. Li, Z.H.; Jia, H.P.; Zhang, Y.Q.; Li, S.G.; Wang, X.; Zhang, H. Efficient parallel optimizations of a high-performance SIFT on GPUs. *J. Parallel Distrib. Comput.* **2018**, *124*, 79–91. [CrossRef]
30. Filippo, B.; Alvise, B.; Francesco, B.; Sandro, C.; Mauro, S. Multi-view horizon-driven sea plane estimation for stereo wave imaging on moving vessels. *Comput. Geosci.* **2016**, *95*, 127–132.
31. Alexis, M.; James, H.; Jason, F.; John, R.; Frederic, D. Incorporating Wave Spectrum Information in Real-time Free-surface Elevation Forecasting: Real-sea Experiments. *IFAC Pap.* **2018**, *51*, 225–230.
32. Zhang, X.X.; Liu, Q.; Liu, J.G.; Zhang, T.; Niz, Y. Binocular Stereo Vision Based Obstacle Detection Method for Manipulator. In Proceedings of the 2016 International Conference on Electrical Engineering and Automation (ICEEA2016), Hangzhou, China, 24–25 April 2016; 439–447.
33. Zhang, J.; Ye, L.; Zhang, Q.; Wang, J.J. Three-Dimensional Object Surface Reconstruction Based on Camera Calibration and SIFT. *Appl. Mech. Mater.* **2015**, *3749*, 719–720. [CrossRef]
34. Jansen, E. Deepwater Drilling Milestone Motion Compensation Platform Enables Offshore Drilling from Vessel. *Sea Technol.* **2020**, *61*, 1712–1715.

applied sciences

MDPI

Article

Fault Diagnosis Method Based on Time Series in Autonomous Unmanned System

Zhuoran Xu [1], Manyi Wang [2,*], Qianmu Li [1] and Linfang Qian [2]

[1] School of Computer Science and Engineering, Nanjing University of Science and Technology, Nanjing 210094, China; xuzhuoran@njust.edu.cn (Z.X.); qianmu@njust.edu.cn (Q.L.)
[2] School of Mechanical Engineering, Nanjing University of Science and Technology, Nanjing 210094, China; lfqian@vip.163.com
* Correspondence: manyi.wang@njust.edu.cn

Abstract: There are various types of autonomous unmanned systems, covering different spaces of sea, land, and air, and they are comprehensively going deep into multiple fields of national security and social life. Due to the development of technology, the scale of unmanned systems is getting larger and larger, the number of components in the system is increasing, and the operating environment of the system is also becoming more and more complex. Therefore, the probability of failure of the components of the system will also be significantly increased. In order to eliminate the impact of the fault in time, the fault diagnosis method is significant. Considering the differences of components in autonomous unmanned systems, if a specific fault diagnosis algorithm is designed for each type of component, it will bring difficulties to the coordinated control of the system. Therefore, this paper analyzes the data characteristics of unmanned autonomous system devices (such as sensors) and finds that these data have time series. Therefore, the data of different devices can be converted into time series, and a general fault diagnosis algorithm suitable for most devices can be studied. The fault diagnosis algorithm is based on the clustering algorithm. In order to improve the clustering effect, the time series of different devices are represented by Gaussian mixture clustering to reduce the computational complexity of the clustering calculation. Then, a time series similarity measurement method based on the improved Markov chain is proposed. This method can better distinguish normal samples from abnormal samples so as to classify and identify faults effectively.

Keywords: autonomous unmanned system; fault diagnosis; clustering algorithm; time series symbolization; improved Markov chain

1. Introduction

The unmanned system consists of a single or multiple unmanned platforms, which can complete designated tasks autonomously or through remote control. The system is highly integrated with mechanization, information, and intelligent platforms to form a smart unmanned system, including a single unmanned aerial vehicle, unmanned vehicles, unmanned ships, unmanned submersibles, and other typical objects.

In order to ensure the reliable execution of the tasks of the multi-unmanned system, it is necessary to pay attention to all aspects of the work of the multi-unmanned system and ensure the safety of each part for overall safety. The autonomous fault monitoring and fault-tolerant control of multiple unmanned systems is a hot research topic at home and abroad. Compared with the traditional single-control system, considering the scale and complexity of the multi-unmanned system, the multi-agent system lacks a central node to coordinate the overall behavior of the system. As a result, the failure of a single node can easily spread to the whole system through the coordination and cooperation between nodes, bringing severe challenges to the system security.

Reference [1] designs a fault estimator that satisfies the hybrid passive/H∞ performance index for stochastic multi-agent systems with unknown actuator offset faults and partial actuator failure faults. Reference [2] proposes a fault-tolerant tracking control design method based on LMI technology and Lyapunov stabilization method, which is applied to aircraft with sensor failures so that for disturbances and sensor failures, the closed-loop system is asymptotically stable at a given disturbance attenuation level. Reference [3] uses Kirsch operator and a deep convolutional neural network to identify abnormal areas of photovoltaic modules, extract defect features, and then use a multi-class support vector machine for defect detection. Reference [4] performs interval sampling and reconstruction of the surface vibration acceleration signals of the rotor UAV under different flight states. It then uses a one-dimensional convolutional neural network (1D-CNN) to perform the reconstruction. Fault identification can effectively identify different or weak fault types with slight differences. Reference [5] applies the transfer learning algorithm based on a convolutional neural network to the fault diagnosis of MEMS inertial sensor of unmanned aerial vehicle, improving the fault classification performance. Reference [6] proposed a credit assignment-based fuzzy cerebellar model articulation controller (FCA-CMAC) neural network information fusion model. The model is used as an estimator for unknown continuous faults, and the proposed fault identification method can diagnose thrusters' continuous, uncertain, and novel failure modes. The heading angle sensor (compass) signal, the yaw rate, and the control signal are used as the input of the FCA-CMAC, and the fault diagnosis results are obtained through offline training. Reference [7] utilizes the characteristics of parallel processing and highly self-organizing and self-learning information of a fuzzy neural network to diagnose the diesel engine of unmanned cabin, overcoming the limitation of a single system and obtaining better fault diagnosis results. Reference [8] proposes a fault diagnosis strategy for induction motors based on support vector machine (SVM) multi-classification. Fault features extracted from electrical and mechanical diagnostic media are used as input to the support vector machine, which performs feature data fusion.

The above study shows that most of the existing fault intelligent diagnosis methods are diagnosis algorithms for a specific target, such as an oil cylinder, partial actuator, etc., and the algorithms are designed using its characteristics (e.g., physical characteristics, electrical characteristics). Therefore, each algorithm has different assumptions and applicable objects and lacks universality. However, there are many types of equipment in autonomous unmanned systems, and their characteristics are also different. If each equipment adopts different fault diagnosis algorithms, it will cause great difficulties for the system's autonomous fault diagnosis and algorithm integration. Therefore, it is important to design a general fault diagnosis algorithm suitable for most equipment.

The completion of autonomous unmanned systems missions is usually based on time, and the sampled equipment data have a strong dependence on time. Time series refers to a dataset in which the data are sorted according to the order of generation time. The recorded data are the monitoring data of the same indicator of the same device and object over time. These data have a trend with time change sequence. Time series is one of the most common forms of data recording, which widely exists in various fields such as underwater ships, aerospace, and so on.

The main contributions of this paper are as follows.

(1) This paper abandons the previous idea of designing a fault diagnosis algorithm based on the characteristics of special equipment and takes the time series characteristics of unmanned autonomous system data as the main research object. Based on the above assumptions, this paper proposes a fault diagnosis method suitable for most equipment.

(2) To improve the applicability of the fault diagnosis algorithm, this paper proposes a fault diagnosis algorithm based on time series. This algorithm and clustering method are combined for fault detection so that autonomous unmanned systems can reveal fault data's inherent laws and properties by learning unlabeled training samples.

(3) In order to reduce the computational complexity of clustering, this paper proposes a time series symbolization method based on Gaussian mixture clustering to express

different types of sensor data uniformly. Through the above processing, the time series can effectively reduce the dimension and calculation.

(4) In order to improve the clustering effect, a time series similarity measurement algorithm based on improved Markov chain is proposed. The algorithm can better extract features and make the discrimination between samples more obvious.

2. Related Work

As a source of information, sensors play an essential role in automatic control systems, and the accuracy of measurement results directly affects the entire system's operation. Faulty sensor signals can lead to incorrect decisions that affect system safety and productivity. Researchers have designed many methods for the fault diagnosis of sensors, such as the diagnosis method based on signal analysis [9–13], and the diagnosis method based on fault pattern recognition [14–18].

The approach adopted in this paper is to treat the sensor data as a time series. Time series classification has been widely used in all aspects of life such as pattern recognition, industrial control, anomaly detection, etc. Time series data have high dimensions and are difficult to classify. Symbolic representation of time series refers to the representation of high-dimensional and noisy continuous real-valued data into low-dimensional and intuitive symbolic sequence data on the premise of maintaining and reflecting the basic characteristics of time series data. There are many studies on the characterization of time series. Keogh et al. presented piecewise approximate polymerization (PAA) [19]. This method is a way to reduce the dimension of time series data, which can effectively improve computing efficiency and reduce the cost of data storage. Lin et al. first proposed SAX, a symbol approximation representation method based on the piecework approximation aggregation method [20]. The value is mapped to the corresponding coincident space according to the segmentation points of the sequence under the Gaussian curve for symbolic representation. Fuad proposes an improved method for traditional symbolic approximate representation to solve the problem of distance measurement. This method uses UMD distance to update the lookup table, making it more advantageous in lower bound tightness and time complexity compared with the original method [21]. Random shifting-based SAX is an improved method proposed by Bai, which can significantly improve the tightness of the lower bound without increasing the corresponding representation granularity [22]. Malinowski put forward a way that adds trend information based on traditional symbol approximation aggregation. The central idea is to use linear regression for each segment to find the mean and slope and convert the time series into a binary string, the fitting error and classification effect are better than the good traditional symbol approximate aggregation method [23]. Yin et al. put forward a new kind of long-time series symbol representation method based on the trends. The difference from other methods is to focus on retaining most of the trend features and patterns of the original series [24]. The above techniques are all based on the assumption that the standardized time series obeys Gaussian distribution, which is often difficult to be established in the natural environment. Especially in autonomous unmanned systems, data do not follow normal distribution due to sampling and other reasons, so the method often leads to large errors in the subsequent process. Pham, N.D. et al. proposed an adaptive time series representation aSAX (adaptive SAX), which combines the traditional symbol approximation representation method with the k-means clustering method to find adaptive symbol segmentation points. This method is superior to the classical time series representation method in all kinds of data sets with or without Gaussian distribution, but the time complexity increases [25]. In 2016, Barnaghi, P.M. proposed an enhanced time series representation method; this method is built on the basis of extended SAX. The maximum, minimum, and average values of each time series are constructed as vectors, the cosine similarity of the vector is used to measure the similarity of the vector, and k-means clustering method is used to divide the symbol region. The calculation time of this method is complicated, and the promotion effect is not significant [26]. Passos, H.D. et al. proposed three different symbolic representation methods, including the k-means-based representation method, k-means-based symbolic aggregation approximate representation method, and k-means-based extended symbolic aggregation approximate representation method [27].

Another way to explore the symbolization of time series is to use it as a text and to learn from the way of thinking of natural language processing. Lin et al. proposed a histogram-based time series representation BOP (bag-of-patterns) based on the bag-of-words model [28]. This method considers the global structure of time series and retains the local structure. This kind of symbol representation method based on natural language processing requires large memory, and it has high time complexity in the training phase, so it cannot be used well in an autonomous unmanned system.

3. Core Idea

3.1. Problem Statement

Time Series: In this paper, time series with the same time interval are abbreviated as:

$$S = \langle s_1, s_2, \ldots, s_n \rangle$$

Time series data describe the changes of various parameters of equipment in work and also contain the law and trend of fault occurrence, which is an important basis for fault detection in this paper.

Cluster-Based Fault Diagnosis: In the fault diagnosis algorithm based on machine learning, it can be divided into two categories: supervised learning and unsupervised learning. Due to the confidentiality of the autonomous unmanned system, the small number of faulty samples, and the difficulty of manual interpretation under massive data, it is impossible to perform supervised learning by effectively labeling the dataset. The learning of unlabeled training samples is needed to reveal the inherent laws and properties of the autonomous unmanned system data. Clustering is often the preferred method for such unsupervised learning.

The operation of dividing samples into a data set into several disjoint subsets is called clustering, and each subset is a cluster. Formally speaking, the sample set $D = \{x_1, x_2, \ldots, x_n\}$ contains n unmarked samples, and for each sample, the goal of the clustering algorithm is to divide the set composed of these samples into several disjoint clusters $C = \{c_i \,|i = 1, 2, \ldots, k\}$, where k is the number of clusters, and $c_{i'} \cap c_i = \varnothing, D = \cup_{i=1}^{k} c_i$.

In order to improve the clustering effect of time series and better discover abnormal data, we need to solve the following problems, denoted as questions Q1–2:

(1) Design a time series dimensionality reduction method to reduce the cost of clustering calculation (Q1);
(2) Determine the similarity measure of two time series (Q2).

3.2. Overview of Our Model

To solve the problems mentioned above, we designed a multi-sensor fault diagnosis method based on time series, as shown in Figure 1.

First, the time series data of the autonomous unmanned system are collected, and the time series is dimensionally reduced by the symbolic representation method. Second, the similarity calculation is performed on the symbolized time series. Third, based on the similarity calculation results, a clustering algorithm is selected to identify abnormal data. In the end, we provide answers to questions Q1–3. The key techniques are as follows:

(1) When extracting and classifying time series, the characteristics of multi-unmanned system time series are first analyzed: high dimensionality and unknown distribution. The high dimensionality determines that we must represent the time series symbolically in order to achieve the purpose of dimensionality reduction. In the choice of the time series symbolization method, this paper proposes a symbolic time series classification method based on Gaussian mixture clustering (in Section 4.1.).

(2) The measurement of time series similarity is the core of cluster-based fault diagnosis algorithms. Whether the two time series are similar mainly depends on whether their changing trends are consistent. However, time series has the characteristics of high dimension and many data types, which brings great inconvenience to the follow-up research. Therefore, for the measurement of time series similarity, the similarity measurement function is particularly important. In Section 4.2., we define a time series similarity measure based on an improved Markov chain.

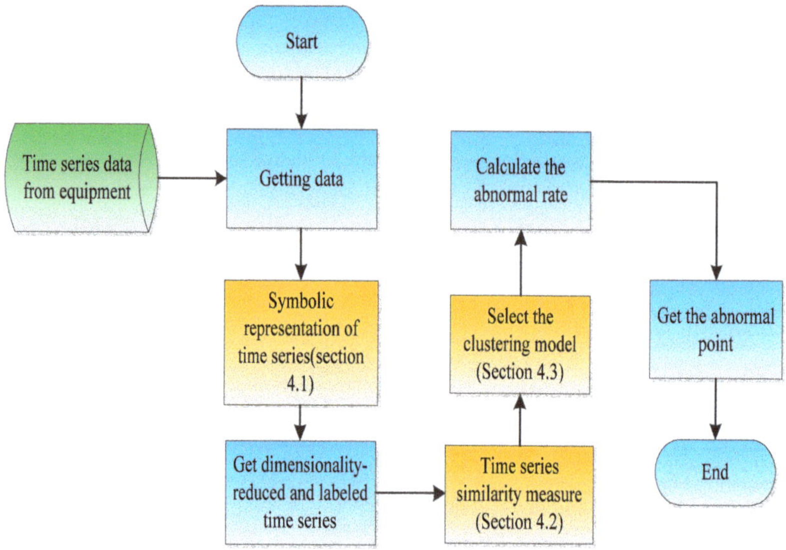

Figure 1. Flow chart.

4. Fault Diagnosis Method Using Improved Clustering Algorithm

4.1. Time Series Symbolic Representation Based on Gaussian Mixture Clustering

For the above problem Q1, considering the distribution characteristics of sensor data, the Gaussian mixture clustering is adopted to represent symbolically. This clustering method is used to select the segmented points of the symbol interval, which has better adaptability to the data with unknown distribution.

The Gaussian mixture model is a probability distribution model with the following form:

$$P_{GMM}(x|\theta) = \sum_{i=1}^{k} \alpha_i \cdot p(x|\theta_i) \tag{1}$$

where $\alpha_i > 0$ is the mixed component of the i-th Gaussian distribution, and $\sum_{i=1}^{k} \alpha_i = 1 p(x|\theta_i)$ is the density of the Gaussian distribution, $\theta_i = (\mu_i, \sigma_i^2)$, where μ_i, σ_i^2 are the location parameters and scale parameters of the i-th Gaussian distribution corresponding to the sub-model are as follows:

$$p(x|\theta_i) = \frac{1}{\sqrt{2\pi}\sigma_i} e^{\left(-\frac{(x-\mu_i)^2}{2\sigma_i^2}\right)} \tag{2}$$

The Gaussian mixture clustering algorithm is a prototype-based clustering method that uses a probability model to describe the clustering prototype. The core idea is to initialize the Gaussian mixture distribution model and then use the EM algorithm to update the model iteratively.

The Algorithm 1 is shown as follows.

Algorithm 1 Gaussian mixture clustering algorithm

Input: sample set $D = \{x_1, x_2, \ldots, x_m\}$, the number of Gaussian mixture components k;
Output: cluster partition $C = \{C_1, C_2, \ldots, C_k\}$;

1. **initialize** the parameter set of Gaussian mixture model $\{(\alpha_i, \mu_i, \sigma_i) | 1 \leq i \leq k\}$
2. **repeat**
3. **for** x_j $(1 \leq j \leq m)$ **do**
4. calculate the posterior probability $E\left(\gamma_{ji}\right)$ generated by the i-th Gaussian distribution

$$E\left(\gamma_{ji}\right) = \frac{\alpha_i \cdot p\left(x_j \middle| \mu_i, \sigma^2{}_i\right)}{\sum_{l=1}^{k} \alpha_i \cdot p\left(x_j \middle| \mu_l, \sigma^2{}_l\right)}$$

5. **end for**
6. **for** i = 1,2,. . . , k do
7. update the mixing coefficient, position parameters and scale parameters for each Gaussian distribution model:

$$\alpha_i^{'} = \frac{\sum_{j=1}^{m} E\left(\gamma_{ji}\right)}{m}; \mu_i^{'} = \frac{\sum_{j=1}^{m} E\left(\gamma_{ji}\right) x_j}{\sum_{j=1}^{m} E\left(\gamma_{ji}\right)}$$

$$\sigma_i^{2'} = \frac{\sum_{j=1}^{m} E\left(\gamma_{ji}\right) \left(x_j - \mu_i^{'}\right)^2}{\sum_{j=1}^{m} E\left(\gamma_{ji}\right)}$$

8. **end for**
9. update the parameter setof Gaussian mixture model $\{(\alpha_i, \mu_i, \sigma_i) | 1 \leq i \leq k\}$ to $\left\{\left(\alpha_i^{'}, \mu_i^{'}, \sigma_i^{'}\right) \middle| 1 \leq i \leq k\right\}$
10. until the end of the iteration
11. **for** j = 1,2,. . . , m do
12. $\lambda_j = arg \max_{i \in \{1,2,\ldots,k\}} \gamma_{ji}$
13. $C_{\lambda_j} = C_{\lambda_j} \cup \left\{x_j\right\}$
14. **end for**

The time series symbolic representation based on Gaussian mixture clustering is to conduct Gaussian mixture clustering on all sample points and use the label obtained by clustering to symbolize the value at a certain time so as to achieve the dimensional reduction of the original time series.

This method can adapt well to the time series data provided by the autonomous unmanned system, which is subject to Gaussian mixture distribution. The model is simple and easy to understand, and the time complexity is relatively low.

4.2. Time Series Feature Engineering Using the Improved Markov Chain Model

The measurement of time series similarity is the core of clustering algorithm. The definition of time series similarity is described as follows:

For the two given time series S_1, S_2, and given a similarity measure function $Dist(S_1, S_2)$. If the time series S_1 and S_2 meet the following conditions:

$$Dist(S_1, S_2) \leq \varepsilon \quad (3)$$

then, the time series S_1 and S_2 are similar, where ε is the threshold of time series similarity.

In the design of the time series similarity measurement method, we must consider the common characteristics of sensor data of unmanned autonomous systems. The ultimate goal of unmanned autonomous system fault diagnosis is to find out the abnormal state of the system. Through analysis, it can be found that the current state of the unmanned autonomous system is determined by the state of the previous moment, independent of other time states, and has Markov properties. Therefore, this paper proposes a similarity measurement method for time series based on Markov chain models.

In Section 4.1, the symbolic representations of the original space mission time series were made through the symbolic representation of time series based on Gaussian mixture clustering, and this symbolic expression can be regarded as different states of the time series, and the process of symbols change can be considered as the state transition of the time series, this transition is determined by the current state, it has the Markov property. Therefore, we regard each time series as a Markov chain, and its state space is the symbol set after the time series is symbolized. Therefore, we designed a time series similarity measurement method based on Markov chain. The specific process is shown in Figure 2.

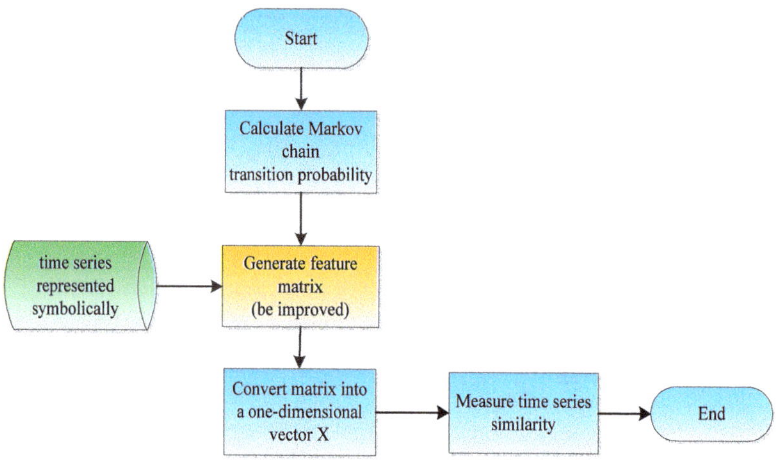

Figure 2. The flow of time series similarity measurement method is based on Markov chain.

Each symbolized time series $S = \{s_1, s_2, \ldots, s_n\}$ satisfies Markov property, that is:

$$P(S_{l+1} = j | S_0 = i_0, S_1 = i_1, \ldots, S_l = i_l) = P(S_{l+1} = j | S_l = i_l) \quad (4)$$

Among them, S_l represents the corresponding state of time series at time l, and the conditional transfer probability $P(\cdot|\cdot)$.

Thus, the transition probability between states in each symbolized time series can be obtained as follows:

$$p_{i,j} = P(S_{l+1} = j | S_l = i) \quad (5)$$

The corresponding transfer probability matrix of each symbolized time series can be obtained, that is:

$$P = \left(p_{i,j}\right) \quad (6)$$

where $\sum_{i,j \in S} p_{i,j} = 1$ and $p_{i,j} \geq 0$.

Since the symbolic representation method of time series is based on Gaussian mixture, clustering gives the symbol set corresponding to the time series, and the length of the symbol set is fixed. We might set the length as k, then the size of the transfer probability matrix corresponding to each time series is also fixed, that is $k \times k$. Therefore, the method can convert the time series with variable length into the transfer probability matrix with the same size, and it also provides effective support for the similarity measure of time series. Because the transition probability matrix is a sparse matrix, the representation of triples is used to reduce the space complexity during storage.

In order to facilitate similarity measurement, the transition probability matrix is converted into a one-dimensional vector X, and the transformation method is as follows:

$$x_{i*k+j} = p_{i,j} \tag{7}$$

Since the one-dimensional vector X can describe each time series S, the similarity of the two time series can be described as the Euclidean distance of the corresponding one-dimensional vector.

Because the time series data will be in a state for quite a long time when a time series is directly modeled using a Markov chain, the transition from one state to its own state will account for the majority. The proportion to other states is much smaller than this value, which leads to this method describing the proportion of different states and ignoring the changes between states. Therefore, this paper improves the time series feature engineering based on Markov chain model.

First, the state transition is divided into two part: maintaining the current state and transitioning to other states. For the part that keeps the current state, the proportion of the state in the whole time series is measured, and the state transition probability after the state transition to other states is measured by removing the state transition to itself. The specific process is as follows:

- Input: Symbolized time series $S = \{s_1, s_2, \ldots, s_m\}$, symbol set C
- Output: feature matrix corresponding to the time series M
- Process: Traverse the symbolized time series S, count the number of transitions between states, and construct a matrix of times as follows:

$$\mathrm{Count}_{i,j} = \sum\nolimits_{S_{l+1}=i \wedge S_{l+1}=j \wedge i,j \in C} 1 \tag{8}$$

Traverse the matrix and calculate the total number of times that the time series maintains its own state, as follows:

$$\mathrm{Sum}_{keep} = \sum\nolimits_{i \in C} \mathrm{Count}_{i,i} \tag{9}$$

Traverse the matrix and calculate the total number of jumps to other states in the time series as follows:

$$\mathrm{Sum}_{change} = \sum\nolimits_{i \in C, j \in C, i \neq j} \mathrm{Count}_{i,j} \tag{10}$$

Use the following formula to obtain the feature matrix corresponding to the time series M

$$M_{i,j} = \begin{cases} \mathrm{Count}_{i,j}/\mathrm{Sum}_{keep} & i \in C, j \in C, i = j \\ \mathrm{Count}_{i,j}/\mathrm{Sum}_{change} & i \in C, j \in C, i \neq j \end{cases} \tag{11}$$

In Section 5.2, we will compare these two different feature extraction methods.

4.3. Fault Diagnosis Method Based on Clustering

There are three types of anomaly detection methods based on clustering:

- The first type of method is based on the assumption that each piece of normal data belongs to a cluster, while the abnormal data do not belong to any cluster. Generally, this type of method does not force every instance to belong to a cluster.
- The second type of method is based on the following assumption: normal data distribution is close to the center of the cluster, while abnormal data distribution is far away from the cluster's center. In this method, the data are firstly clustered, then the anomaly is evaluated by calculating the distance from each point to the center of the corresponding cluster.
- The third method is based on the following assumptions: the samples of the cluster where the normal data are located are relatively dense, while the samples of the cluster where the abnormal data are located are relatively sparse.

Considering that the abnormal data of autonomous unmanned systems belong to the small sample data, the fault diagnosis can choose the first and third types of clustering algorithms. The first type of method mainly uses DBSCAN clustering method based on density, while the third type is based on k-means based on partition and hierarchical clustering. In this paper, we adopt the third clustering method for fault detection.

5. Experimental Results and Analysis

In order to verify the validity of the method, we carried out further experiments by using the data set of an aeronautical autonomous unmanned system.

5.1. Model Evaluation Criteria

(1) Data fitting effect evaluation index

This paper uses the fitting error to measure the effect of time series coincidence. The segmentation model of time series is the set of elements composed of the start time and the end time of the subsequence, which is denoted as $M = \{(b_0, e_0), (b_1, e_1), \ldots, (b_k, e_k)\}$. The element (b_0, e_0) is a two-tuple, in which b_i is the start time of the segmentation subsequence, and e_i is the end time of the segmentation subsequence.

The symbolization of time series can be regarded as the segmentation and description of time series. For time series $S = (x_1, t_1), (x_2, t_2), \ldots, (x_n, t_n)$, the segmentation mode M is adopted, and the fitting error is:

$$E = \sqrt{\sum_{j=0}^{k} \sum_{i=b_j}^{e_j} \left(x_i - avg_j\right)^2} \tag{12}$$

where avg_j is the mean value of the i-th subsequence.

Considering that the selected time series are not equally long, to measure the effect of time series symbolization better, we add a normalization factor to the fitting error to balance the length of time series, that is:

$$E\prime = \frac{1}{n}\sqrt{\sum_{j=0}^{k} \sum_{i=b_j}^{e_j} \left(x_i - avg_j\right)^2} \tag{13}$$

The smaller the fitting error, the better the effect of time series coincidence.

(2) Clustering effect evaluation index

Considering the data confidentiality and labeling difficulties, we use internal indexes of clustering to evaluate the clustering effect. In order to better describe the index, some function definitions are given.

Function definition 1: Average distance between samples in the cluster:

$$avg(C) = \frac{2}{|C|(|C|-1)} \sum_{1 \le i \le j \le |C|} dist\left(x_i, y_j\right)$$

Function definition 2: Maximum distance between samples in the cluster:

$$diam(C) = \max_{1 \le i \le j \le |C|} dist\left(x_i, y_j\right)$$

Function definition 3: The distance between the nearest samples between clusters:

$$d_{min}(C_i, C_j) = \min_{x_i \in C_i, y_j \in C_j} dist\left(x_i, y_j\right)$$

Function definition 4: Distance of sample center point between clusters

$$d_{cen}(C_i, C_j) = dist\left(\mu_i, \mu_j\right)$$

For cluster partition $C = \{C_1, C_2, \ldots, C_k\}$, there are several common internal indexes of clustering performance measurement as follows:

Internal index 1: The Davies–Bouldin Index (DBI)

$$DBI = \frac{1}{k}\sum_{i=1}^{k} \max_{j \neq i}\left(\frac{avg(C_i) + avg(C_j)}{d_{cen}(C_i, C_j)}\right) \quad (14)$$

Internal index 2: Dunn Index (DI)

$$DI = \min_{1 \leq i \leq k}\left\{\min_{j \neq i}\left(\frac{d_{min}(C_i, C_j)}{\max_{1 \leq l \leq k} diam(C_l)}\right)\right\} \quad (15)$$

For clustering, if the average value between samples of each cluster is smaller, the DB index is smaller. If the distance between the cluster center points is larger, the DB index is smaller. In contrast, the Dunn exponent characterizes the ratio between the minimum of the closest distance between any two clusters and the maximum of the distance between the two furthest points within any cluster. If the minimum value of the nearest distance between any two clusters is larger, the Dunn index is larger; if the maximum value of the distance between the two furthest points in any cluster is smaller, the Dunn index is larger.

5.2. Test Results and Analysis

(1) Experiment results of time series symbolization

Each value of all time series is taken as a sample to train the Gaussian mixture clustering model, and the clustering label of each sample is successively obtained as its corresponding symbol thus each time series is symbolized, as shown in the figure below: The series is symbolized, as shown in Figure 3 below:

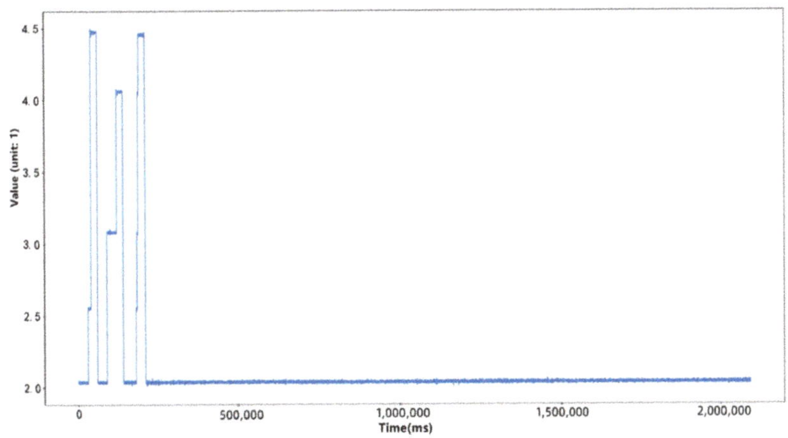

Figure 3. Original time series.

Figures 4 and 5 respectively compress a time series in the data set with the traditional symbolic representation method of time series and the symbolic representation method of time series based on Gaussian mixture clustering proposed in this paper. These two algorithms use the same compression ratio, and it can be seen that the symbolic representation method of time series based on Gaussian mixture clustering proposed in this paper can better describe the fluctuation of the original time series.

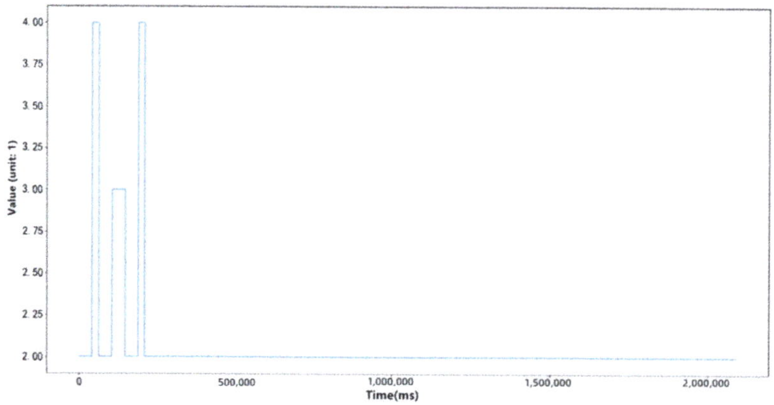

Figure 4. Time series after SAX dimensionality reduction compression.

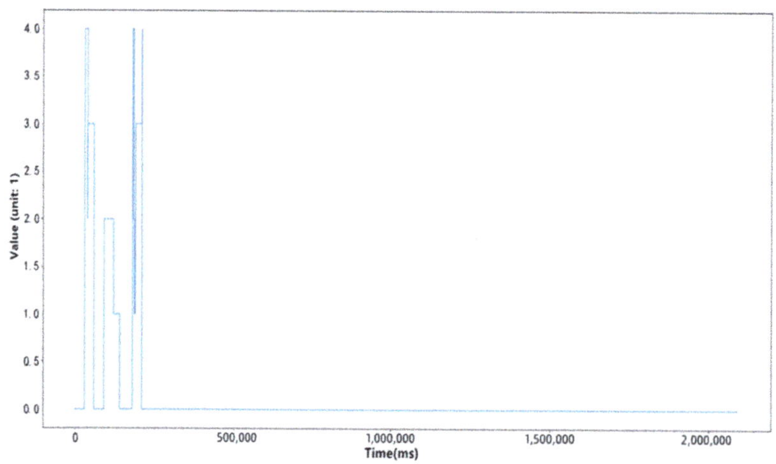

Figure 5. Time series of SAX dimensionality reduction compression based on gaussian mixture clustering.

We used the above formula to calculate 408 time series in the data set for experiments, and the results are shown in Table 1.

Table 1. Comparison between the symbolic representation method of time series based on Gaussian mixture clustering and the traditional method.

The Serial Number	Traditional Symbolic Representation of Time Series	A Symbolic Representation Method of Time Series Based on Gaussian Mixture Clustering
1	0.006521201	0.002199259
2	0.002446704	0.015722783
3	8.02×10^{-7}	8.02×10^{-7}
4	0.00701202	0.000718564
5	0.002896344	0.002896344
6	0.002338207	0.009036553
7	0.005271108	0.009092318
8	0.004497003	0.001726509

Table 1. *Cont.*

The Serial Number	Traditional Symbolic Representation of Time Series	A Symbolic Representation Method of Time Series Based on Gaussian Mixture Clustering
9	0	0
10	0.004288419	0.009207164
...	...	...
471	5.84×10^{-7}	5.84×10^{-7}
472	0.009355902	0.001267874
473	3.12×10^{-7}	3.12×10^{-7}
474	0.008298582	0.003469096
475	0.004413301	0.008241421
476	0.010358657	0.02383774
477	0.011526121	0.025266455
478	8.52×10^{-7}	8.52×10^{-7}
479	0.009612222	0.001176151
480	0.008940187	0.002242722
The average error	0.003751851	0.003683018

As can be seen from the above table, the fitting error of the time series symbolized representation method based on Gaussian mixture clustering is smaller than that of the traditional time series symbolized representation method. From this perspective, this method can describe the original time series relatively well.

(2) Clustering experiment results based on improved similarity measurement method

According to Section 4.2., an improved Markov chain model is used to model the time series. For example, all of the time-series data are converted into a 10×10 state transition matrix and converted into a vector of length 100, as described in this chapter. Based on using the principal component analysis to reduce data dimension for the vector visualization, Figure 6 shows the sample distribution based on the Markov chain feature engineering, Figure 7 shows the sample distribution after based on the improved Markov chain feature engineering (A blue dot in Figures 6 and 7 represents a data sample).

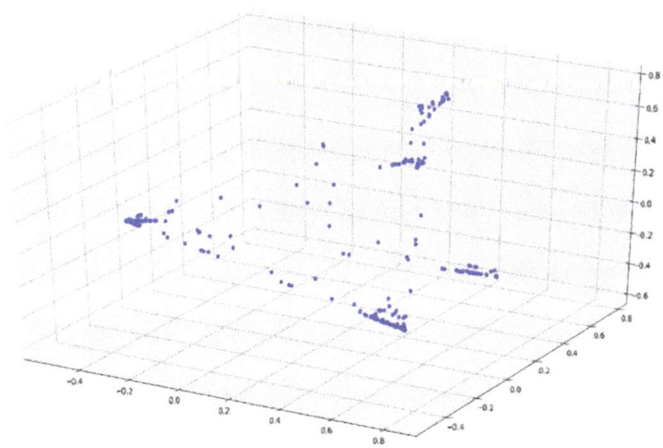

Figure 6. Sample distribution based on the Markov chain feature engineering.

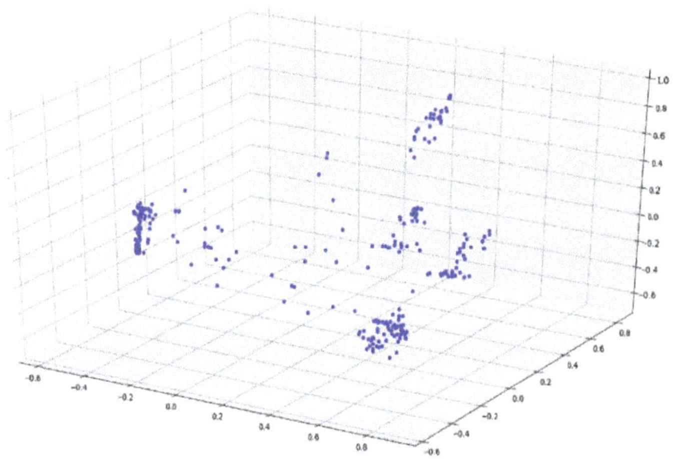

Figure 7. Sample distribution based on the improved Markov chain feature engineering.

It can be observed that the discrimination between the samples based on the improved Markov chain feature engineering is more prominent, which is conducive to the following clustering analysis.

The DBI indices of different clustering algorithms based on the Markov feature engineering and the improved Markov feature engineering are obtained by calculation, as shown in Figure 8:

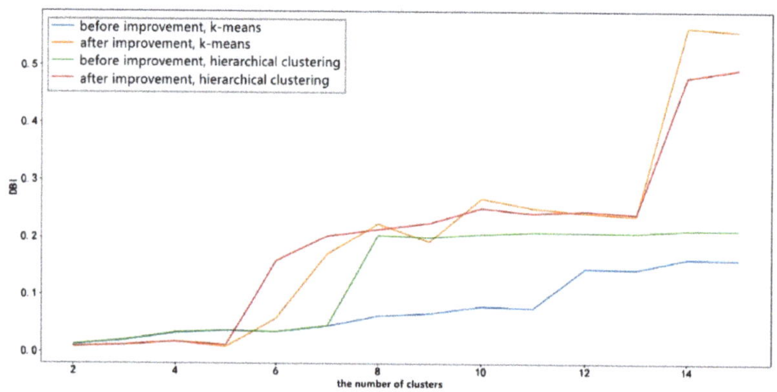

Figure 8. Comparison diagram of clustering DBI index before and after improvement.

In general, when the number of clusters exceeds 5, the DBI of the improved Markov feature engineering under any clustering method is much smaller than the DBI before the improvement. The effect is more apparent when the number of clusters increases.

The DI of different clustering algorithms based on the Markov feature engineering and the improved Markov feature engineering is obtained by calculation, as shown in Figure 9.

It can be found from the figure that when the number of clusters is between 2 and 7, the DI increases, but after that, the DI starts to decrease. When the number of clusters is greater than 11, the DI of the improved feature engineering is still relatively stable. In contrast, the DI of the pre-improvement feature engineering is very small under any algorithm, and the clustering effect becomes generally poor.

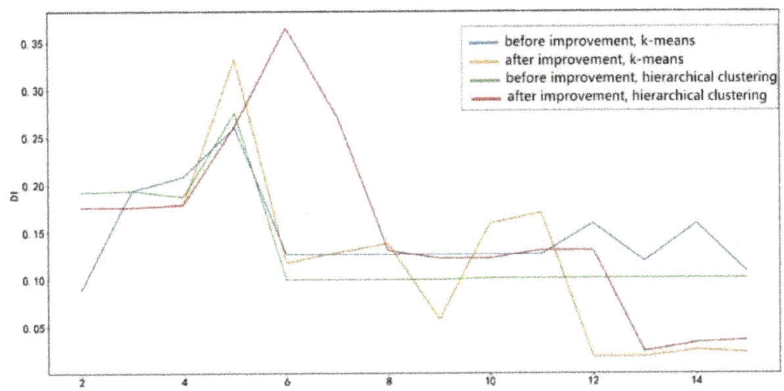

Figure 9. Comparison diagram of clustering DI index before and after improvement.

(3) Abnormal sample detection results

Two example of fault diagnosis based on the improved k-means clustering algorithm are shown in Figures 10 and 11 (Different colors are used to represent different clusters in the diagram).

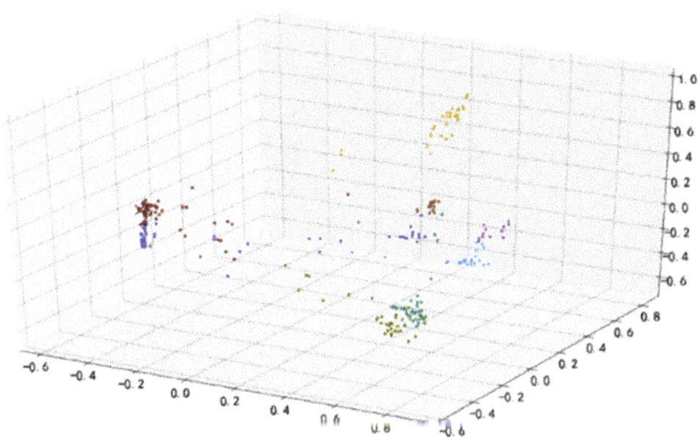

Figure 10. Cluster distribution under k-means (normal sample).

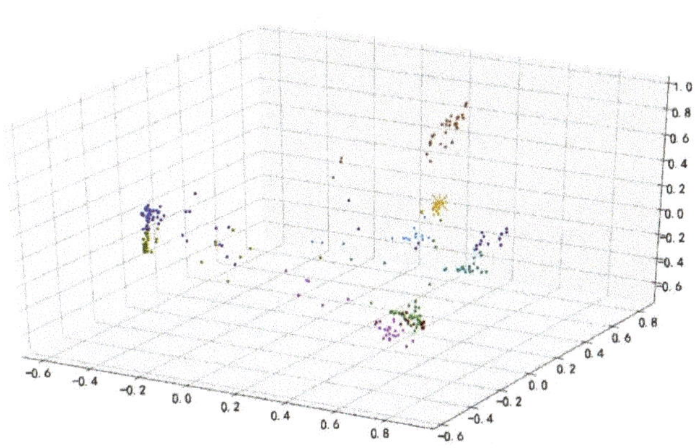

Figure 11. Cluster distribution under k-means (abnormal sample).

The sample anomaly rate is defined as follows in this paper:

$$ACC(S) = \frac{n_S}{N} \tag{16}$$

N is the total number of samples, n_S is the number of elements in the cluster where the current sample is located. In order to compare the accuracy of samples and enhance the robustness of the algorithm, a normalization factor is introduced to the normality of samples

$$ACC'(S) = \lambda \frac{n_S}{N} \tag{17}$$

λ is the normalization factor that is used to normalize the function, which the following formula can obtain:

$$\lambda = \frac{ACC(S) - \min_{i \in D} ACC(i)}{\max_{i \in D} ACC(i) - \min_{i \in D} ACC(i)} \tag{18}$$

Then, the abnormal rate of the sample is:

$$P(S) = 1 - ACC'(S) \tag{19}$$

The experiment result (Figure 10) shows a normal sample using the improved clustering algorithm. In this case, there are 10 clusters, and the cluster in which the sample is located is number 2. No. 2 cluster has 55 samples. The total number of samples is 480, the smallest cluster has 10 samples, and the largest cluster has 73 samples. The abnormal rate of this sample is 28.75%.

The experiment result (Figure 11) shows an abnormal sample using the improved clustering algorithm. In this case, there are 10 clusters, and the cluster in which the sample is located is number 5. No. 5 cluster has 17 samples. The total number of samples is 481, the smallest cluster has 10 samples, and the largest cluster has 73 samples. The abnormal rate of this sample is 88.89%.

The larger the value is, the fewer elements in the cluster, and the greater the possibility of abnormality.

6. Conclusions

In the research of fault detection algorithm of the autonomous unmanned system, most researchers focus on the use of fault detection algorithm of a specific device and the lack of a unified fault detection method for multiple devices in the system. The specific work is as follows:

- Based on the time series data of unmanned autonomous system and the time series data of various devices, a fault diagnosis algorithm based on time series and clustering is proposed. This method can be applied to a variety of devices.
- A time series symbolic representation method based on Gaussian mixture clustering is proposed. By symbolizing the time series to perform dimensionality reduction operations on the data, it can reduce the complexity of clustering calculations and reduce the impact of noise data on autonomous data. Compared with other time series symbolization methods, this method has better fitting effect and is more convenient for subsequent clustering operations.
- A method for measuring the similarity of time series based on Markov chain model is proposed. This method further improves the original method and performs different probability calculations for the transition between states and the transition of the state itself to further optimize the clustering effect.
- Through experiments, we found that the improved clustering algorithm can better detect abnormal data.

Author Contributions: Conceptualization, Z.X. and Q.L.; methodology, Z.X. and M.W.; writing—original draft preparation, Z.X.; project administration, L.Q. All authors have read and agreed to the published version of the manuscript.

Funding: This work was supported in part by The 4th project "Research on the Key Technology of Endogenous Security Switches" (2020YFB1804604) of the National Key R&D Program "New Network Equipment Based on Independent Programmable Chips" (2020YFB1804600), the 2020 Industrial Internet Innovation and Development Project from Ministry of Industry and Information Technology of China, the Fundamental Research Fund for the Central Universities (30918012204, 30920041112), the 2019 Industrial Internet Innovation and Development Project from Ministry of Industry and Information Technology of China.

Data Availability Statement: The data presented in this study are available from the corresponding authors. The data cannot be made public as it relates to ongoing projects.

Conflicts of Interest: We declare that we have no financial and personal relationships with other people or organizations that can inappropriately influence our work, and there is no professional or other personal interest of any nature or kind in any product, service and/or company that could be construed as influencing the position presented in, or the review of this manuscript.

References

1. Liu, M.; Ke, Y.; Wang, A.; Liu, Z.; Li, J. Fault-tolerant control approach for multi-agent systems with multiple failures. *J. Hangzhou Dianzi Univ. (Nat. Sci.)* **2022**, *42*, 7.
2. Qian, M.; Xiong, K. Fault tolerant control scheme design for formation flight control system of multiple unmanned aerial vehicles. *Trans. Nanjing Univ. Aeronaut. Astronaut.* **2018**, *35*, 693–701.
3. Li, X.; Li, W.; Yang, Q.; Yan, W.; Zomaya, A.Y. An unmanned inspection system for multiple defects detection in photovoltaic plants. *IEEE J. Photovolt.* **2020**, *10*, 568–576. [CrossRef]
4. Du, C.; Zhang, X.; Zhong, R.; Li, F.; Yu, F.; Rong, Y.; Gong, Y. Unmanned aerial vehicle rotor fault diagnosis based on interval sampling reconstruction of vibration signals and a one-dimensional convolutional neural network deep learning method. *Meas. Sci. Technol.* **2022**, *33*, 065003. [CrossRef]
5. Gao, T.; Sheng, W.; Yin, Y.; Du, X. A transfer learning based unmanned aerial vehicle mems inertial sensors fault diagnosis method. *J. Phys. Conf. Ser.* **2021**, *1852*, 042084. [CrossRef]
6. Zhu, D.; Sun, B. Information fusion fault diagnosis method for unmanned underwater vehicle thrusters. *IET Electr. Syst. Transp.* **2013**, *3*, 102–111. [CrossRef]
7. Yuan, X.; Peng, J.; Yang, D.; Hao, R.S. Research on automatic monitoring and fault diagnosis system of unmanned cabin. *EPH Int. J. Sci. Eng.* **2017**, *3*, 1–7.
8. Martínez-Morales, J.D.; Palacios-Hernández, E.R.; Campos-Delgado, D.U. Multiple-fault diagnosis in induction motors through support vector machine classification at variable operating conditions. *Electr. Eng.* **2018**, *100*, 59–73. [CrossRef]
9. Cheng, G.; Chen, X.H.; Shan, X.L.; Liu, H.G.; Zhou, C.F. A new method of gear fault diagnosis in strong noise based on multi-sensor information fusion. *J. Vib. Control* **2016**, *22*, 1504–1515. [CrossRef]
10. Othman, M.S.; Nuawi, M.; Mohamed, R. Vibration and acoustic emission signal monitoring for detection of induction motor bearing fault. *Int. J. Eng. Res. Technol.* **2015**, *4*, 924–929.
11. Seshadrinath, J.; Singh, B.; Panigrahi, B.K. Investigation of vibration signatures for multiple fault diagnosis in variable frequency drives using complex wavelets. *IEEE Trans. Power Electron.* **2013**, *29*, 936–945. [CrossRef]
12. Estima, J.O.; Cardoso, A.J.M. A new algorithm for real-time multiple open-circuit fault diagnosis in voltage-fed pwm motor drives by the reference current errors. *IEEE Trans. Ind. Electron.* **2013**, *60*, 3496–3505. [CrossRef]
13. Martinez-Herrera, A.L.; Ferrucho-Alvarez, E.R.; Ledesma-Carrillo, L.M.; Mata-Chavez, R.I.; Lopez-Ramirez, M.; Cabal-Yepez, E. Multiple fault detection in induction motors through homogeneity and kurtosis computation. *Energies* **2022**, *15*, 1541. [CrossRef]
14. Liu, Z.; Guo, W.; Tang, Z.; Chen, Y. Multi-sensor data fusion using a relevance vector machine based on an ant colony for gearbox fault detection. *Sensors* **2015**, *15*, 21857–21875. [CrossRef] [PubMed]
15. Xiong, G.; Shi, D.; Chen, J.; Lin, Z.; Duan, X. Divisional fault diagnosis of large-scale power systems based on radial basis function neural network and fuzzy integral. *Electr. Power Syst. Res.* **2013**, *105*, 9–19. [CrossRef]
16. Yu, C.; Ma, S.; Wen, X. Application of natural gradient algorithm for the aircraft engine vibration signal separation and fault diagnosis. *J. Converg. Inf. Technol.* **2012**, *7*, 382–388.
17. Li, P.; Chai, Y.; Cen, M.; Qiu, Y.; Zhang, K. Multiple fault diagnosis of analog circuit using quantum hopfield neural network. In Proceedings of the 25th Chinese Control and Decision Conference (CCDC), Guiyang, China, 25–27 May 2013; pp. 4238–4243.
18. Guo, X.-G.; Tian, M.-E.; Li, Q.; Ahn, C.K.; Yang, Y.-H. Multiple-fault diagnosis for spacecraft attitude control systems using rbfnn-based observers. *Aerosp. Sci. Technol.* **2020**, *106*, 106195. [CrossRef]
19. Keogh, E.; Chakrabarti, K.; Pazzani, M.; Mehrotra, S. Dimensionality reduction for fast similarity search in large time series databases. *Knowl. Inf. Syst.* **2001**, *3*, 263–286. [CrossRef]

20. Lin, J.; Keogh, E.; Lonardi, S.; Chiu, B. A symbolic representation of time series, with implications for streaming algorithms. In Proceedings of the 8th ACM SIGMOD Workshop on Research Issues in Data Mining and Knowledge Discovery, San Diego, CA, USA, 13 June 2003; Association for Computing Machinery: San Diego, CA, USA, 2003; pp. 2–11.
21. Muhammad Fuad, M.M. Modifying the symbolic aggregate approximation method to capture segment trend information. In *Modeling Decisions for Artificial Intelligence*; Torra, V., Narukawa, Y., Nin, J., Agell, N., Eds.; Springer International Publishing: Cham, Switzerland; pp. 230–239.
22. Bai, X.; Xiong, Y.; Zhu, Y.; Zhu, H. Time series representation: A random shifting perspective. In Proceedings of the 14th International Conference on Web-Age Information Management, Beidaihe, China, 14–16 June 2013; Springer: Beidaihe, China, 2013; pp. 37–50.
23. Malinowski, S.; Guyet, T.; Quiniou, R.; Tavenard, R. 1d-sax: A novel symbolic representation for time series. In *Advances in Intelligent Data Analysis XII*; Tucker, A., Höppner, F., Siebes, A., Swift, S., Eds.; Springer: Berlin/Heidelberg, Germany; pp. 273–284.
24. Yin, H.; Yang, S.-Q.; Zhu, X.-Q.; Ma, S.-D.; Zhang, L.-M. Symbolic representation based on trend features for knowledge discovery in long time series. *Front. Inf. Technol. Electron. Eng.* **2015**, *16*, 744–758. [CrossRef]
25. Pham, N.D.; Le, Q.L.; Dang, T.K. Two novel adaptive symbolic representations for similarity search in time series databases. In Proceedings of the 2010 12th International Asia-Pacific Web Conference, Buscan, Korea, 6–8 April 2010; pp. 181–187.
26. Barnaghi, P.M.; Bakar, A.A.; Othman, Z.A. Enhanced symbolic aggregate approximation method for financial time series data representation. In Proceedings of the 2012 6th International Conference on New Trends in Information Science, Service Science and Data Mining (ISSDM2012), Taipei, Taiwan, 23–25 October 2012; pp. 790–795.
27. dos Santos Passos, H.; Teodoro, F.G.S.; Duru, B.M.; de Oliveira, E.L.; Peres, S.M.; Lima, C.A.M. Symbolic representations of time series applied to biometric recognition based on ecg signals. In Proceedings of the 2017 International Joint Conference on Neural Networks (IJCNN), Anchorage, AL, USA, 14–19 May 2017; pp. 3199–3207.
28. Lin, J.; Khade, R.; Li, Y. Rotation-invariant similarity in time series using bag-of-patterns representation. *J. Intell. Inf. Syst.* **2012**, *39*, 287–315. [CrossRef]

applied sciences

MDPI

Article

In-Motion Coarse Alignment Method Based on Position Loci and Optimal-REQUEST for SINS

Haoqian Huang * and Jiaying Wei

College of Energy and Electrical Engineering, Hohai University, Nanjing 211100, China; xtcwjy@hhu.edu.cn
* Correspondence: hqhuang@hhu.edu.cn

Abstract: In this paper, an improved in-motion coarse alignment method is proposed for a strapdown inertial navigation system (SINS) using position loci obtained from the Global Positioning System (GPS). The difference from the popular coarse alignment methods is that the proposed algorithm uses GPS position loci information to form the vector observation, and does not need velocity information, which expands the application range of in-motion coarse alignment. In addition, this paper utilizes the Optimal-REQUEST algorithm to reduce the influence of random errors contained in the vector observation. The Optimal-REQUEST algorithm is an adaptive iterative updating algorithm, which can adaptively adjust the gain of the filter according to the loss function. Simulation results confirmed that the proposed algorithm can suppress the impact of random errors effectively. The pitch, roll and yaw angles calculated by the proposed algorithm were improved by 51.95%, 53.80% and 63.03% compared with the comparison algorithms.

Keywords: coarse alignment; strapdown inertial navigation system; attitude estimation; observation vector

1. Introduction

A strapdown inertial navigation system (SINS) is an autonomous navigation system that does not rely on any external information. A SINS has the characteristics of excellent concealment because it does not radiate energy to the outside. Thus, it can work in a variety of complex environments, such as air, ground, and underwater [1–3]. In practical navigation applications, an integrated navigation system (INS) is usually used to improve navigation accuracy. An INS usually takes a SINS as the main navigation system and the Global Positioning System (GPS) or other systems, where errors do not accumulate with time, are used as auxiliary navigation systems [4–6]. Among many INSs, SINS/GPSs have attracted the attention of many researchers because of their wide application range and high efficiency [7]. In SINS/GPS research, initial alignment has been one of the significant focal areas [8].

Initial alignment is the process of calculating the initial attitude, position, and velocity of vehicles [9,10]. In a SINS/GPS integrated system, the initial position and velocity can be easily obtained by GPS. Therefore, the essence of initial alignment is to calculate the initial attitude. In general, initial alignment is divided into two stages: coarse alignment and fine alignment [11]. The coarse alignment period estimates and provides a rough attitude matrix for the fine alignment. Then, the fine alignment uses the liner model and filter technology based on the results of coarse alignment. Thus, the accuracy of coarse alignment greatly affects the accuracy of the whole integrated navigation process [12–14].

Several researchers have devoted themselves to devising new coarse alignment algorithms. In [15–17], the coarse alignment algorithms are studied based on gravitational apparent motion. The authors of [15] used gravitational apparent motion vectors at three moments to construct an attitude matrix and then to obtain latitude information by calculating the geometric relationship among the apparent motions. However, the method proposed in [15] is easily affected by random noise contained in accelerometer measurement. However, when the measurement time interval is very short, or the measurement quantity is inaccurate, the apparent motion of gravity will be collinear. In order to avoid collinearity among apparent motions, the authors of [16] propose an online sensor data denoising method and study the design of a novel reconstruction method for apparent motion Through the reconstruction of vectors, the method can increase the accuracy of self-alignment. In addition, in [17], a fast self-alignment method with real-time estimated latitude to obtain high accuracy with unknown geographic latitude constraints and limited time is proposed. Unfortunately, the above three coarse alignment methods can only be applied to the static base or swaying base and limit the application range of a SINS. Therefore, researchers have begun to investigate in-motion coarse alignment methods based on an external equipment auxiliary.

When a vehicle is moving, the coarse alignment process cannot be completed using measurement information from inertial sensors alone, and it is necessary to engage the assistance of external equipment. GPS is the most popular auxiliary equipment for land vehicles, and many coarse alignment methods based on GPS have been studied in recent years. The authors of [18] investigated an optimization-based alignment (OBA) algorithm in an innovative way. The OBA algorithm uses infinite vector observations to transform the coarse alignment problem into an attitude determination problem. In [19], the approach reported in [18] is further developed by utilization of devised velocity integration formulae to construct the vector observation. Based on the above two algorithms, the basic problem of in-motion alignment can be solved. Unfortunately, both algorithms cannot suppress the impact of outliers contained in the outputs of GPS and inertial measurement units. Thus, in [20], a robust optimization-based alignment method to detect the outliers and improve the accuracy of in-motion coarse alignment is proposed. It is worth noting that the above in-motion coarse alignment methods are all based on the ground velocity. However, not all GPS receivers can export the ground velocity and some of them can only provide the position information of the vehicle. This limits the application of in-motion coarse alignment. Therefore, ref. [21] investigated an in-motion coarse alignment method for the SINS/GPS integrated system using position loci. The vector observation, based on position trajectory, further expands the application range of coarse alignment on the moving base. However, ref. [21] uses a filter-QUEST algorithm to calculate the attitude matrix, which has only limited ability to reduce the errors of vector observation contained in the inertial sensors and GPS measurement. Furthermore, the propagation noises are treated as fixed fading memory values in the filter-QUEST algorithm, which is similar to the REQUEST algorithm, and this makes both of the algorithms suboptimal.

To improve the performance of in-motion coarse alignment, this paper extends the research of [21] and proposes a new alignment method based on an Optimal-REQUEST algorithm and position loci (OP-PL). This paper employs a vector observation calculation method based on position loci. Then, the Optimal-REQUEST is utilized to suppress the noise in the measurement vectors and to filter the measurement noises by adaptively adjusting the fading memory value. Finally, the objective of improving the accuracy of coarse alignment is achieved. The contributions of the proposed in-motion coarse alignment are summarized as follows:

1. The proposed OP-PL utilizes position information to construct the observation vectors, which can expand the application range of the in-motion coarse alignment algorithm, because the proposed method only utilizes the position information which is available from every GPS receiver.
2. The outliers contained in the GPS positioning outputs can be suppressed as the outliers can be smoothed by GPS position loci; this can reduce the noise impact of outliers on observation vectors.
3. The proposed algorithm employs the Optimal-REQUEST algorithm to further suppress random noise contained in the observation vectors. The Optimal-REQUEST overcomes the defect of the REQUEST algorithm of the use of an empirical and con-

stant gain to filter the random noise. Therefore, the Optimal-REQUEST algorithm can optimally filter random noise using a cost function, and can improve the stability and accuracy of the coarse alignment.

The remainder of the paper is organized as follows: Section 2 introduces the traditional method of in-motion coarse alignment and defines the problem. Next, Section 3 elaborates the principle of vector observation based on the position loci, and Section 4 introduces the proposed Optimal-REQUEST algorithm. The simulation experiments conducted are described in Section 5. Finally, the conclusions and suggestions for future work are presented in Section 6. See Appendix A for the proof of some formulas.

2. Traditional Method of In-Motion Coarse Alignment and Related Work

In this section, the traditional method of in-motion coarse alignment is briefly described. Then, the limits and practical issues of constructing a vector using velocity measurement are introduced.

2.1. Traditional Construction Method Using Velocity Measurement

The core issue for in-motion coarse alignment is to find the initial attitude matrix C_b^n. According to the chain rule of the attitude matrix, the initial attitude matrix C_b^n can be divided into three matrices [11]:

$$C_b^n(t) = C_{n(0)}^{n(t)} C_{b(0)}^{n(0)} C_{b(t)}^{b(0)} \tag{1}$$

where n-frame denotes the navigation frame aligned with East-North-Up (ENU) geodetic axes, and b-frame denotes the body frame aligned with inertial measurement unit axes, namely right-forward-upper. $C_{n(t)}^{n(0)}$ and $C_{b(t)}^{b(0)}$ encode the attitude changes of the navigation frame and the body frame from the initial time to the current time, respectively. Using the attitude matrix differential equation, $C_{n(t)}^{n(0)}$ and $C_{b(t)}^{b(0)}$ can be calculated as follows:

$$\dot{C}_{n(t)}^{n(0)} = C_{n(t)}^{n(0)} \omega_{in}^n \times \tag{2}$$

$$\dot{C}_{b(t)}^{b(0)} = C_{b(t)}^{b(0)} \omega_{ib}^b \times \tag{3}$$

where i-frame denotes the Earth-center-fixed orthogonal reference frame, which does not rotate with the alignment process, ω_{in}^n denotes the rotation of n-frame relative to the i-frame, and ω_{ib}^b denotes the rotation of the b-frame relative to the i-frame. After the above calculation, the determination of C_b^n is translated into the determination of $C_{b(0)}^{n(0)}$, which is a constant matrix during the in-motion coarse alignment.

2.2. Related Work

To calculate the matrix $C_{b(0)}^{n(0)}$, ref. [18] proposes an algorithm using vector observation. Using the traditional vector observation method, we can acquire the measurement vectors and the reference vectors based on the velocity measurement. The specific force equation under the navigation frame is given by [18]:

$$\dot{v}^n = C_b^n f^b - (2\omega_{ie}^n + \omega_{en}^n) \times v^n + g^n \tag{4}$$

Combining Equations (1)–(3), the following equation can be obtained:

$$C_{b(0)}^{n(0)} C_{b(t)}^{b(0)} f^b = C_{n(t)}^{n(0)} \dot{v}^n + C_{n(t)}^{n(0)} (2\omega_{ie}^n + \omega_{en}^n) \times v^n - C_{n(t)}^{n(0)} g^n \tag{5}$$

Then, the reference vectors and the measurement vectors can be acquired, and the initial alignment problem is transformed into an attitude determination problem.

$$C_{b(0)}^{n(0)}\alpha_v = \beta_v \tag{6}$$

where

$$\alpha_v = \int_0^t C_{b(\tau)}^{b(0)} f^b d\tau \tag{7}$$

$$\beta_v = C_{n(t)}^{n(0)} v^n - v^n(0) + \int_0^t C_{n(\tau)}^{n(0)} \omega_{ie}^n \times v^n d\tau - \int_0^t C_{n(\tau)}^{n(0)} g^n d\tau \tag{8}$$

From Equation (8), it can be seen that the measurement vector β_v is calculated based on the ground velocity v^n; this requires that the GPS receiver can output the ground velocity. This construction method for the observation vector is the method reported in many current alignment studies [12–14]. However, some GPS receivers can only output the position of the vehicle, and the traditional construction method of vector observation is not suitable for in-motion coarse alignment in this situation. Furthermore, the noise contained in the GPS and inertial sensor measurements inevitably affects the accuracy of observation. Thus, it is necessary to change the construction method of vector observation and expand the application scope of the traditional OBA algorithms.

3. The Principle of Vector Observation Based on Position Loci

In this section, a construction method of vector observation based on position loci is derived, and the detailed recursive formulae are also investigated.

3.1. Position Integration Formula

Based on the principles of inertial navigation, the position and the velocity under the navigation frame are shown as:

$$\dot{p}^n = R_c v^n \tag{9}$$

where $p^n = \begin{bmatrix} L & \lambda & h \end{bmatrix}^T$; L, λ, h denote the vehicle's latitude, longitude, and height, respectively. The local curvature matrix R_c is defined below.

$$R_c = \begin{bmatrix} 0 & \frac{1}{R_M+h} & 0 \\ \frac{\sec L}{R_N+h} & 0 & 0 \\ 0 & 0 & 1 \end{bmatrix} \tag{10}$$

where R_M and R_N denote the meridian radius and the transverse radius of curvature based on the WGS-84 reference ellipsoid, respectively.

As is well-known, the expression of position integration can be obtained by integrating the velocity. Combining with Equation (9), the specific position integration can be calculated as:

$$r^n = \int_0^t v^n d\tau = \int_0^t R_c^{-1} \dot{p}^n d\tau \tag{11}$$

where $r^n(0) = 0$.

Based on the relation between position and velocity, Equations (6)–(8) can be reconstructed through integral calculation, as follows:

$$C_{b0}^{n0} \int_0^t \alpha_v d\tau = \int_0^t \beta_v d\tau \tag{12}$$

The integral calculation formulae on both sides of the equation are shown below [22]:

$$\alpha_p = \int_0^t \alpha_v d\tau = \int_0^t \int_0^\tau C_{b(\sigma)}^{b0} f^b d\sigma d\tau \tag{13}$$

$$\beta_p = \int_0^t C_{n(\tau)}^{n0}\dot{r}^n d\tau - tv^n(0) + \int_0^t\int_0^\tau C_{n(\sigma)}^{n0}(\omega_{ie}^n\times)\dot{r}^n d\sigma d\tau - \int_0^t\int_0^\tau C_{n(\sigma)}^{n0} g^n d\sigma d\tau \quad (14)$$

Comparing Equation (14) with Equation (8), we notice that the measurement vectors are constructed by the GPS position information of the vehicle. Moreover, the effects of outliers, coming from the GPS position information, can be weakened through position-loci integration. Thus, the proposed algorithm is more robust than the traditional algorithms.

3.2. Discrete Recursive Formulae

In the previous subsection, we obtained the vector construction method based on position information. However, all the formulae are continuous and cannot be calculated in practical experiments. Thus, the discrete recursive formulae are studied in this subsection [21].

The reference vector α_p can be approximated as below:

$$\alpha_p(M) = \sum_{k=0}^{M-1}\Delta t_G \sum_{m=0}^{k-1} C_{b(t_m)}^{b0}\int_{t_m}^{t_{m+1}} C_{b(\sigma)}^{b(t_m)} f^b d\sigma + \sum_{k=0}^{M-1} C_{b(t_k)}^{b0}\int_{t_k}^{t_{k+1}}\int_{t_k}^{\tau} C_{b(\sigma)}^{b(t_k)} f^b d\sigma d\tau \quad (15)$$

where Δt_G and M represent the sampling time of GPS and the current time, respectively. Specifically, $t = t_M = M\Delta t_G$, $t_k = k\Delta t_G$ and $t_m = m\Delta t_G$. The integral term can be calculated by the two-sample method, which is as follows:

$$\int_{t_m}^{t_{m+1}} C_{b(\sigma)}^{b(t_m)} f^b d\sigma = \Delta v_1 + \Delta v_2 + \frac{1}{2}(\Delta\theta_1 + \Delta\theta_2)\times(\Delta v_1 + \Delta v_2) + \frac{2}{3}(\Delta\theta_1\times\Delta v_2 + \Delta v_1\times\Delta\theta_2) \quad (16)$$

$$\int_{t_k}^{t_{k+1}}\int_{t_k}^{\tau} C_{b(\sigma)}^{b(t_k)} f^b d\sigma d\tau = \frac{\Delta t_G}{30}(25\Delta v_1 + 5\Delta v_2 + 12\Delta\theta_1\times\Delta v_1 + 8\Delta\theta_1\times\Delta v_2 + 2\Delta v_1\times\Delta\theta_1 + 2\Delta\theta_2\times\Delta v_2) \quad (17)$$

where

$$\begin{cases} \Delta v_1 = \sum_{l=1}^{\frac{D}{2}} f_l^b \Delta t_S & \Delta v_2 = \sum_{l=\frac{D}{2}+1}^{D} f_l^b\Delta t_S \\ \Delta\theta_1 = \sum_{l=1}^{\frac{D}{2}} \omega_{ib,l}^b\Delta t_S & \Delta\theta_2 = \sum_{l=\frac{D}{2}+1}^{D}\omega_{ib,l}^b\Delta t_S \end{cases} \quad (18)$$

where D denotes the half sampling time of GPS, and Δt_s denotes the sampling time of the accelerometers and gyroscopes.

After acquiring the recursive formulae of the reference vector, the measurement vector is derived in detail. The first integration term of Equation (14) is shown below:

$$\int_0^t C_{n(\tau)}^{n0}\dot{r}^n d\tau = C_{n(t)}^{n0} r^n - \int_0^t C_{n(\tau)}^{n0}(\omega_{in}^n\times) r^n d\tau = C_{n(t_M)}^{n0} r^n - \sum_{k=0}^{M-1} C_{n(t_k)}^{n0}\int_{t_k}^{t_{k+1}} C_{n(\tau)}^{n(t_k)}(\omega_{in}^n\times) r^n d\tau \quad (19)$$

The second integration term can be given by:

$$\begin{aligned} &\int_0^t\int_0^\tau C_{n(\sigma)}^{n0}(\omega_{ie}^n\times)\dot{r}^n d\sigma d\tau = \int_0^t C_{n(\tau)}^{n0}(\omega_{ie}^n\times) r^n d\tau - \int_0^t\int_0^\tau C_{n(\sigma)}^{n0}(\omega_{in}^n\times)(\omega_{ie}^n\times) r^n d\sigma d\tau \\ &= \sum_{k=0}^{M-1} C_{n(t_k)}^{n0}\int_{t_k}^{t_{k+1}} C_{n(\tau)}^{n(t_k)}(\omega_{ie}^n\times) r^n d\tau - \sum_{k=0}^{M-1}\Delta t_G\sum_{m=0}^{k-1} C_{n(t_m)}^{n0}\int_{t_m}^{t_{m+1}} C_{n(\tau)}^{n(t_m)}(\omega_{in}^n\times)(\omega_{ie}^n\times) r^n d\tau \\ &- \sum_{k=0}^{M-1} C_{n(t_k)}^{n0}\int_{t_k}^{t_{k+1}}\int_{t_k}^{\tau} C_{n(\sigma)}^{n(t_k)}(\omega_{in}^n\times)(\omega_{ie}^n\times) r^n d\sigma d\tau \end{aligned} \quad (20)$$

The last integration term is calculated as follows:

$$
\begin{aligned}
&\int_0^t \int_0^\tau C_{n(\sigma)}^{n0} g^n d\sigma d\tau = \sum_{k=0}^{M-1} \int_{t_k}^{t_{k+1}} \int_0^\tau C_{n(\sigma)}^{n0} g^n d\sigma d\tau = \sum_{k=0}^{M-1} \int_{t_k}^{t_{k+1}} \left(\int_0^{t_k} C_{n(\sigma)}^{n0} g^n d\sigma + \int_{t_k}^{\tau} C_{n(\sigma)}^{n0} g^n d\sigma \right) d\tau \\
&= \sum_{k=0}^{M-1} \Delta t_G \int_0^{t_k} C_{n(\sigma)}^{n0} g^n d\sigma + \sum_{k=0}^{M-1} \int_{t_k}^{t_{k+1}} \int_{t_k}^{\tau} C_{n(\sigma)}^{n0} g^n d\sigma d\tau \\
&= \sum_{k=0}^{M-1} \Delta t_G \sum_{m=0}^{k-1} C_{n(t_m)}^{n0} \int_{t_m}^{t_{m+1}} C_{n(\sigma)}^{n(t_m)} g^n d\sigma + \sum_{k=0}^{M-1} C_{n(t_k)}^{n0} \int_{t_k}^{t_{k+1}} \int_{t_k}^{\tau} C_{n(\sigma)}^{n(t_k)} g^n d\sigma d\tau
\end{aligned} \tag{21}
$$

Substituting Equations (19)–(21) into Equation (14), the complete formula for the measurement vector is shown below:

$$
\begin{aligned}
\beta_p = {} & C_{n(t_M)}^{n0} r^n - \sum_{k=0}^{M-1} C_{n(t_k)}^{n0} \int_{t_k}^{t_{k+1}} C_{n(\tau)}^{n(t_k)} (\omega_{in}^n \times) r^n d\tau - t v^n(0) + \sum_{k=0}^{M-1} C_{n(t_k)}^{n0} \int_{t_k}^{t_{k+1}} C_{n(\tau)}^{n(t_k)} (\omega_{ie}^n \times) r^n d\tau \\
& - \sum_{k=0}^{M-1} \Delta t_G \sum_{m=0}^{k-1} C_{n(t_m)}^{n0} \int_{t_m}^{t_{m+1}} C_{n(\tau)}^{n(t_m)} (\omega_{in}^n \times)(\omega_{ie}^n \times) r^n d\tau - \sum_{k=0}^{M-1} C_{n(t_k)}^{n0} \int_{t_k}^{t_{k+1}} \int_{t_k}^{\tau} C_{n(\sigma)}^{n(t_k)} (\omega_{in}^n \times)(\omega_{ie}^n \times) r^n d\sigma d\tau \\
& - \sum_{k=0}^{M-1} \Delta t_G \sum_{m=0}^{k-1} C_{n(t_m)}^{0} \int_{t_m}^{t_{m+1}} C_{n(\sigma)}^{n(t_m)} g^n d\sigma - \sum_{k=0}^{M-1} C_{n(t_k)}^{0} \int_{t_k}^{t_{k+1}} \int_{t_k}^{\tau} C_{n(\sigma)}^{n(t_k)} g^n d\sigma d\tau
\end{aligned} \tag{22}
$$

Assuming that the terms related to $(\omega_{ie}^n \times)$ and $(\omega_{in}^n \times)$ are so small compared with other terms in Equation (22) that they can be ignored during calculation, then the measurement vectors can be approximately determined as:

$$
\beta_p = C_{n(t_M)}^{n0} r^n - t v^n(0) - \sum_{k=0}^{M-1} \Delta t_G \sum_{m=0}^{k-1} C_{n(t_m)}^{n0} \int_{t_m}^{t_{m+1}} C_{n(\sigma)}^{n(t_m)} g^n d\sigma - \sum_{k=0}^{M-1} C_{n(t_k)}^{n0} \int_{t_k}^{t_{k+1}} \int_{t_k}^{\tau} C_{n(\sigma)}^{n(t_k)} g^n d\sigma d\tau \tag{23}
$$

Meanwhile, the attitude matrix $C_{n(t)}^{n(t_k)}$ can be determined approximately as follows:

$$
C_{n(t)}^{n(t_k)} = I + \frac{\sin\|\varphi_n\|}{\|\varphi_n\|}(\varphi_n \times) + \frac{1 - \cos\|\varphi_n\|}{\|\varphi_n\|^2}(\varphi_n \times)^2 \approx I + (\varphi_n \times) \tag{24}
$$

where $\varphi_n \approx \int_{t_k}^{t} \omega_{ie}^n d\tau \approx (t - t_k)\omega_{in}^n$. The reason for this approximation is that ω_{in}^n slowly changes during the coarse alignment. Further discretizing the Equation (23), we obtain the following formulae.

$$
\begin{aligned}
\int_{t_m}^{t_{m+1}} C_{n(\sigma)}^{n(t_m)} g^n d\sigma &\approx \int_{t_m}^{t_{m+1}} [I + (\varphi_n \times)] g^n d\sigma = \int_{t_m}^{t_{m+1}} [I + (\sigma - t_m)(\omega_{in}^n \times)] g^n d\sigma \\
&\approx \left(\Delta t_G I + \tfrac{\Delta t_G^2}{2} (\omega_{in}^n \times) \right) g^n
\end{aligned} \tag{25}
$$

$$
\begin{aligned}
\int_{t_k}^{t_{k+1}} \int_{t_k}^{\tau} C_{n(\sigma)}^{n(t_k)} g^n d\sigma d\tau &\approx \int_{t_k}^{t_{k+1}} \int_{t_k}^{\tau} [I + (\varphi_n \times)] g^n d\sigma d\tau = \int_{t_k}^{t_{k+1}} \int_{t_k}^{\tau} [I + (\sigma - t_m)(\omega_{in}^n \times)] g^n d\sigma d\tau \\
&\approx \left(\tfrac{\Delta t_G^2}{2} I + \tfrac{\Delta t_G^3}{6} (\omega_{in}^n \times) \right) g^n
\end{aligned} \tag{26}
$$

Finally, we acquire the recursive calculation process for the observation vector based on the formulae described:

$$
\beta_p \approx C_{n(t_M)}^{n0} r^n - t v^n(0) - \sum_{k=0}^{M-1} \Delta t_G \sum_{m=0}^{k-1} C_{n(t_m)}^{n0} \left(\Delta t_G I + \frac{\Delta t_G^2}{2} (\omega_{in}^n \times) \right) g^n - \sum_{k=0}^{M-1} C_{n(t_k)}^{n0} \left(\frac{\Delta t_G^2}{2} I + \frac{\Delta t_G^3}{6} (\omega_{in}^n \times) \right) g^n \tag{27}
$$

$$
r^n \approx \sum_{k=1}^{M} R_c^{-1}(k)(p_k^n - p_{k-1}^n) \tag{28}
$$

where p_k^n represents the position information of the vehicle at time instant k, and $R_c^{-1}(k)$ is the inverse matrix of Equation (10). It is worth noting that the position information can be acquired through GPS directly.

4. Attitude Determination Using the Optimal-REQUEST Algorithm

Based on the above calculations, models of the observation vector are acquired. However, the results of coarse alignment are inevitably affected by GPS measurement errors in the process of constructing vectors. Thus, an Optimal-REQUEST algorithm, based on position loci (OR-PL), is proposed to reduce the impact of errors and to determine the initial attitude matrix $C_{b(0)}^{n(0)}$. The Optimal-REQUEST algorithm utilizes the measurement vectors and reference vectors to construct a K-matrix. Moreover, an iteration algorithm, which can adjust the gain according to the measurement error, is applied to update the K-matrix. The optimal attitude quaternion can then be obtained by calculating the K-matrix eigenvector corresponding to its largest eigenvalue.

4.1. Measurement Update

When the measurement vectors and reference vectors are obtained based on the position loci, the initial attitude matrix $C_{b(0)}^{n(0)}$ can be determined by minimizing the following formula:

$$\frac{1}{2}\sum_{i=1}^{n} a_i \|b_i - Ar_i\|^2 = 1 - q^T K q \tag{29}$$

where $b_i = \frac{\beta_p}{\|\beta_p\|}$, $r_i = \frac{\alpha_p}{\|\alpha_p\|}$ and $\sum_{i=1}^{n} a_i = 1$. The matrix A is the initial attitude matrix $C_{b(0)}^{n(0)}$. The construction method of the K-matrix is shown below:

$$K_{k/k} = \begin{bmatrix} S_k - \sigma_k I_3 & z_k \\ z_k^T & \sigma_k \end{bmatrix} \tag{30}$$

where the parameters in Equation (30) are defined as follows:

$$B_k \triangleq \sum_{i=1}^{n} a_i b_i r_i^T,\ S_k \triangleq B_k + B_k^T \tag{31}$$

$$z_k \triangleq \sum_{i=1}^{n} a_i b_i \times r_i,\ \sigma_k \triangleq tr(B_k) \tag{32}$$

According to the REQUEST algorithm, the propagation formula of the K-matrix is given by:

$$K_{k+1/k} = \Phi_k K_{k/k} \Phi_k^T \tag{33}$$

During the coarse alignment, the initial attitude matrix $C_{b(0)}^{n(0)}$ stays constant. Thus, the propagation coefficient matrix $\Phi_k = I_4$.

When the new observation vectors are obtained at time $k+1$, we construct the corresponding incremental K-matrix, namely δK_{k+1}. The specific construction method is the same as Equations (30)–(32), and the only difference is that the vectors acquired at time $k+1$ are used to calculate δK_{k+1}.

Next, the estimated $K_{k+1/k+1}$ can be updated by the combination of prior estimation $K_{k+1/k}$ and the incremental matrix δK_{k+1} at time $k+1$, namely,

$$K_{k+1/k+1} = (1-\rho_{k+1})\frac{m_k}{m_{k+1}}K_{k+1/k} + \rho_{k+1}\frac{\delta m_{k+1}}{m_{k+1}}\delta K_{k+1} \tag{34}$$

where m_k, m_{k+1} and δm_{k+1} are scalar coefficients, and the specific relationship of the three scalars is as follows:

$$m_{k+1} = (1 - \rho_{k+1})m_k + \rho_{k+1}\delta m_{k+1} \tag{35}$$

The fading memory factor ρ_{k+1} is the core of the REQUEST algorithm and the Optimal-REQUEST algorithm. The REQUEST algorithm uses the fading memory concept to treat the measurement noise, and this makes the algorithm suboptimal. However, the proposed OP-PL algorithm can optimally filter the measurement noise by calculating the value of the fading memory factor adaptively.

4.2. Stochastic Models

To deduce the error models of the proposed OP-PL algorithm accurately, we assume that the reference vector r_{k+1} is error-free and that the measurement vector b_{k+1} is corrupted by the noisy vector $\delta\mathrm{b}_{k+1}$.

$$\mathrm{b}_{k+1} = \mathrm{A}_{k+1}\mathrm{r}_{k+1} + \delta\mathrm{b}_{k+1} \tag{36}$$

Define the following noise matrix:

$$\mathrm{V}_{k+1} = \begin{bmatrix} \mathrm{S}_b - \sigma_b\mathrm{I}_3 & \mathrm{z}_b \\ \mathrm{z}_b^{\mathrm{T}} & \sigma_b \end{bmatrix} \tag{37}$$

where the parameters in Equation (37) are defined as:

$$\mathrm{B}_b \triangleq \sum_{i=1}^{n} a_i\delta\mathrm{b}_{k+1}(\mathrm{r}_{k+1})^{\mathrm{T}},\ \mathrm{S}_b \triangleq \mathrm{B}_b + \mathrm{B}_b^{\mathrm{T}} \tag{38}$$

$$\mathrm{z}_b \triangleq \sum_{i=1}^{n} a_i\delta\mathrm{b}_{k+1} \times \mathrm{r}_{k+1},\ \sigma_b \triangleq tr(\mathrm{B}_b) \tag{39}$$

It is worth noting that V_{k+1} is made up of $\delta\mathrm{b}_{k+1}$ and r_{k+1}, and $\delta\mathrm{b}_{k+1}$ is random. Thus, we conclude that V_{k+1} is also random. The linear relation of V_{k+1} in $\delta\mathrm{b}_{k+1}$ is used to deduce a stochastic model for V_{k+1} and $\delta\mathrm{b}_{k+1}$. The relationship between $\delta\mathrm{K}_{k+1}$ and V_{k+1} is given by:

$$\delta\mathrm{K}_{k+1} = \delta\mathrm{K}_{k+1}^{o} + \mathrm{V}_{k+1} \tag{40}$$

where $\delta\mathrm{K}_{k+1}^{o}$ denotes the error-free matrix.

The mean and the covariance of $\delta\mathrm{b}_k$ are constructed as follows:

$$E[\delta\mathrm{b}_k] = 0 \tag{41}$$

$$E\left[\delta\mathrm{b}_k\delta\mathrm{b}_{k+i}^{\mathrm{T}}\right] = \mu_k(\mathrm{I}_3 - \mathrm{b}_k\mathrm{b}_{k+i}^{\mathrm{T}})\delta_{k,k+i} \tag{42}$$

where $\delta_{k,k+i}$ represents the Kronecker delta function and μ_k represents the variance of the component of $\delta\mathrm{b}_k$. It is obvious that $\delta\mathrm{b}_k$ is white noise with zero-mean. Thus, V_k also has the characteristics of zero-mean white noise, namely,

$$E\left[\mathrm{V}_k\mathrm{V}_{k+i}^{\mathrm{T}}\right] = 0 \tag{43}$$

where $i \neq 0$. To further derive the uncertainty of the matrix, we introduce a special form of uncertainty, namely P_{xx}. The proof of P_{xx} is shown in Appendix A.

The P-matrix of V_k is given by:

$$\Re_k = E\left[\mathrm{V}_k\mathrm{V}_k^{\mathrm{T}}\right] = \begin{bmatrix} \Re_{11} & \Re_{12} \\ \Re_{21} & \Re_{22} \end{bmatrix} \tag{44}$$

The calculation formulae of each element in the matrix are shown below and the detailed derivation process is shown in [23].

$$\Re_{11} = \frac{\mu_k}{n_k}\sum_{i=1}^{n_k}\left[3-(\mathrm{r}_i^{\mathrm{T}}\mathrm{b}_i)^2\right]\mathrm{I}_3 + (\mathrm{b}_i^{\mathrm{T}}\mathrm{r}_i)(\mathrm{b}_i\mathrm{r}_i^{\mathrm{T}} + \mathrm{r}_i\mathrm{b}_i^{\mathrm{T}}) + (\mathrm{r}_i\times)(\mathrm{b}_i\mathrm{b}_i^{\mathrm{T}})(\mathrm{r}_i\times)^{\mathrm{T}} \tag{45}$$

$$\Re_{12} = 0 \tag{46}$$

$$\Re_{21} = 0^{\mathrm{T}} \tag{47}$$

$$\Re_{22} = \frac{2\mu_k}{n_k} \tag{48}$$

where μ_k is the variance for all the n_k simultaneous observations calculated at time t_k.

4.3. Calculation of Optimal Gain

After the measurement update formulae and error models are deduced, the cost function can be determined through the estimated errors of the K-matrix. Next, the fading memory factor ρ_{k+1} can be updated adaptively according to the cost function.

We define the estimation errors of the K-matrix as:

$$\Delta\mathrm{K}_{k+1/k} \triangleq \mathrm{K}^o_{k+1/k} - \mathrm{K}_{k+1/k} \tag{49}$$

$$\Delta\mathrm{K}_{k+1/k+1} \triangleq \mathrm{K}^o_{k+1/k+1} - \mathrm{K}_{k+1/k+1} \tag{50}$$

where $\mathrm{K}^o_{k+1/k}$ and $\mathrm{K}^o_{k+1/k+1}$ denote the a priori and the a posteriori estimation with error-free, respectively. $\Delta\mathrm{K}_{k+1/k}$ and $\Delta\mathrm{K}_{k+1/k+1}$ denote the errors of the a priori and the a posteriori estimation.

Similarly to Equation (34), reconstruct the equation with the error-free K-matrix, namely:

$$\mathrm{K}^o_{k+1/k+1} = (1-\rho_{k+1})\frac{m_k}{m_{k+1}}\mathrm{K}^o_{k+1/k} + \rho_{k+1}\frac{\delta m_{k+1}}{m_{k+1}}\delta\mathrm{K}^o_{k+1} \tag{51}$$

After subtracting Equations (34) and (51), substitute Equations (40), (49) and (50) into the following formula:

$$\Delta\mathrm{K}_{k+1/k+1} = (1-\rho_{k+1})\frac{m_k}{m_{k+1}}\Delta\mathrm{K}_{k+1/k} + \rho_{k+1}\frac{\delta m_{k+1}}{m_{k+1}}\mathrm{V}_{k+1} \tag{52}$$

The P matrix of $\Delta\mathrm{K}_{k+1/k}$ and $\Delta\mathrm{K}_{k+1/k+1}$ are defined as below

$$\mathrm{P}_{k+1/k} \triangleq E\left[\Delta\mathrm{K}_{k+1/k}\Delta\mathrm{K}^{\mathrm{T}}_{k+1/k}\right] \tag{53}$$

$$\mathrm{P}_{k+1/k+1} \triangleq E\left[\Delta\mathrm{K}_{k+1/k+1}\Delta\mathrm{K}^{\mathrm{T}}_{k+1/k+1}\right] \tag{54}$$

Considering Equation (52), the following formula can be calculated as:

$$\begin{aligned}\Delta\mathrm{K}_{k+1/k+1}\Delta\mathrm{K}^{\mathrm{T}}_{k+1/k+1} &= \left[(1-\rho_{k+1})\tfrac{m_k}{m_{k+1}}\right]^2\Delta\mathrm{K}_{k+1/k}\Delta\mathrm{K}^{\mathrm{T}}_{k+1/k} \\ -(1-\rho_{k+1})\rho_{k+1}\tfrac{m_k\delta m_{k+1}}{m^2_{k+1}}&(\Delta\mathrm{K}_{k+1/k}\mathrm{V}^{\mathrm{T}}_{k+1} + \mathrm{V}_{k+1}\Delta\mathrm{K}^{\mathrm{T}}_{k+1/k}) + (\rho_{k+1}\tfrac{\delta m_{k+1}}{m_{k+1}})^2\mathrm{V}_{k+1}\mathrm{V}^{\mathrm{T}}_{k+1}\end{aligned} \tag{55}$$

As mentioned, $\delta\mathrm{b}_k$ and V_k represent the white noise process with zero-mean. Moreover, $\Delta\mathrm{K}_{k+1/k}$ consists of V_i, where i is an integer from 1 to k. Thus, $\Delta\mathrm{K}_{k+1/k}$ has no relation with V_{k+1}, namely,

$$E\left[\Delta\mathrm{K}_{k+1/k}\mathrm{V}^{\mathrm{T}}_{k+1}\right] = 0 \tag{56}$$

Then, we take the expectation of Equation (55), and simplify the following formulae:

$$P_{k+1/k+1} \triangleq E\left[\Delta K_{k+1/k+1}\Delta K_{k+1/k+1}^{T}\right] = \left[(1-\rho_{k+1})\frac{m_k}{m_{k+1}}\right]^2 E\left[\Delta K_{k+1/k}\Delta K_{k+1/k}^{T}\right] + (\rho_{k+1}\frac{\delta m_{k+1}}{m_{k+1}})^2 E\left[V_{k+1}V_{k+1}^{T}\right] \quad (57)$$

$$P_{k+1/k+1} = \left[(1-\rho_{k+1})\frac{m_k}{m_{k+1}}\right]^2 P_{k+1/k} + (\rho_{k+1}\frac{\delta m_{k+1}}{m_{k+1}})^2 \Re_{k+1} \quad (58)$$

Based on Equation (58), we define the cost function J_{k+1}:

$$J_{k+1} = tr(E\left[\Delta K_{k+1/k+1}\Delta K_{k+1/k+1}^{T}\right]) = tr(P_{k+1/k+1}) \quad (59)$$

The calculation of the fading memory factor is then transformed by solving the minimization problem of Equation (59). We use the derivation to obtain the expression of the fading memory factor.

$$J_{k+1}(\rho_{k+1}) = \left[(1-\rho_{k+1})\frac{m_k}{m_{k+1}}\right]^2 tr(P_{k+1/k}) + (\rho_{k+1}\frac{\delta m_{k+1}}{m_{k+1}})^2 tr(\Re_{k+1}) \quad (60)$$

$$\frac{dJ_{k+1}}{d\rho_{k+1}} = 2\left[\left(\frac{m_k}{m_{k+1}}\right)^2 tr(P_{k+1/k}) + \left(\frac{\delta m_{k+1}}{m_{k+1}}\right)^2 tr(\Re_{k+1})\right]\rho_{k+1} - 2\left(\frac{m_k}{m_{k+1}}\right)^2 tr(P_{k+1/k}) = 0 \quad (61)$$

Finally, the stationary point ρ_{k+1}^* can be determined using the extremum of J_{k+1}.

$$\rho_{k+1}^* = \frac{m_k^2 tr(P_{k/k})}{m_k^2 tr(P_{k/k}) + \delta m_{k+1}^2 tr(\Re_{k+1})} \quad (62)$$

4.4. Algorithm Summary

Compared with the REQUEST algorithm, the OP-PL algorithm uses the updated estimation error to define the special cost function and calculates the scalar gain ρ_{k+1}^*. Thus, the OP-PL can better reduce the impact of GPS measurement errors which are contained in the observation vectors. For clear explanation, the whole in-motion coarse alignment procedure is shown in Table 1 and Figure 1.

Table 1. In-motion coarse alignment based on the OP-PL algorithm.

Initialization	$k = 0$, $m_0 = \delta m_0$, $\rho_0^* = 1$, $C_{n(0)}^{n0} = C_{b(0)}^{b0} = I_3$, $K_{0/0} = \delta K_0$, $P_{0/0} = \Re_0$
Step 1	$k = k + 1$
Step 2	Update $C_{n(t)}^{n(0)}$ and $C_{b(t)}^{b(0)}$ using Equations (2) and (3)
Step 3	Calculate and normalize α_p using Equations (15)–(18)
Step 4	Calculate and normalize β_p using Equations (27) and (28)
Step 5	Calculate δK_k and $\Re_k$ using Equations (30)–(32) and (44)–(48)
Step 6	Calculate ρ_k^* and m_k using Equations (62) and (35)
Step 7	Update $K_{k/k}$ using Equation (34)
Step 8	Calculate the attitude matrix $C_{b(0)}^{n(0)}$
Step 9	Calculate the attitude matrix $C_{b(k)}^{n(k)}$
Step 10	Back to the Step 1 until the end

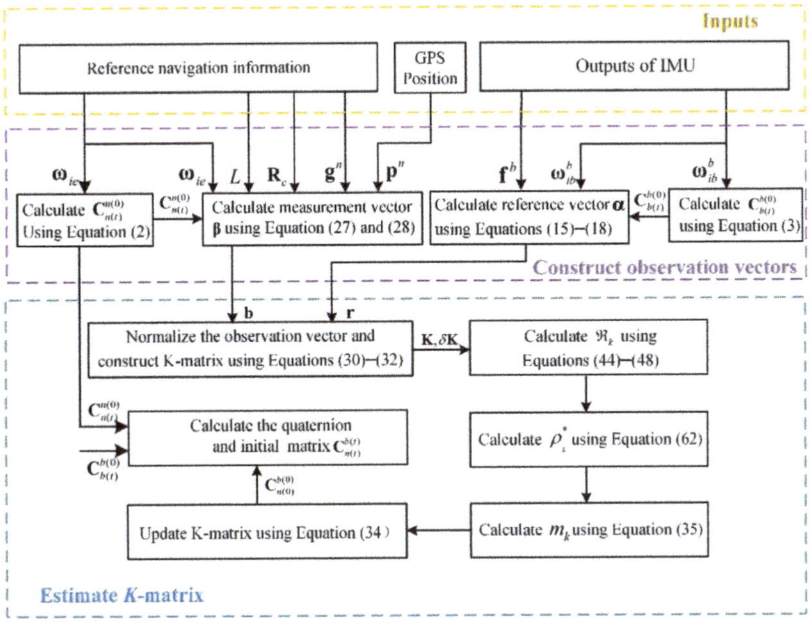

Figure 1. In-motion coarse alignment procedure.

5. Simulation

The simulation tests carried out to verify the accuracy and rapidity of the proposed OP-PL algorithm are described in this section. The trajectory of the vehicle was designed to follow a zigzag movement, and the actual information of the vehicle was known. Thus, the actual angles were used as reference angles to calculate the errors of three angles. The zigzag trajectory and the reference information are shown in Figures 2 and 3, respectively. It can be seen that the outliers, which corrupt the measurement of GPS, were added to the measurement. The output frequencies of GPS and the simulated inertial measurement unit (IMU) were set as 5 Hz and 200 Hz, respectively. The constant bias and random noise of the gyroscopes and accelerometers are listed in Table 2. The initial latitude and longitude were set as 32.00° and 118.00°, respectively. Throughout the in-motion coarse alignment process, it was assumed that the vehicle was moving on the horizontal plane; the initial values of the attitude angles were all set as 0°. The simulation was conducted using Matlab 2016b with an alignment time of 300 s. The proposed OP-PL algorithm was compared with the QUEST algorithm and the REQUEST algorithm to verify performance.

Figures 4–6 show the different performances of the REQUEST algorithm and the proposed OP-PL algorithm. We chose three different values of the fading memory factor, which were 0.1, 0.01 and 0.001, to compare with the optimal adaptive gain ρ^*. Figure 7 shows the optimal gain of the proposed algorithm in the simulation. Considering the results from three angles, we found that the REQUEST algorithm was unstable during simulation. From Figures 4 and 5, it is evident that the REQUEST algorithm showed similar performance to the proposed algorithm when 0.1 was chosen as the value of the fading memory factor. The 'MEAN' and 'RMSE' represent the mean and the root mean square error of the attitude errors. The MEAN and RMSE errors of the blue and black curves were calculated by collecting data from 150 s to 300 s, where the blue curve represents $\rho = 0.1$ and the black curve represents the optimal adaptively gain. The pitch MEAN errors of the blue and black curves were 0.0269° and −0.0760°, respectively. The pitch RMSE errors of the blue and black curves were 0.0425° and 0.0824°, respectively. Although the accuracy of the pitch angle calculated by the proposed OP-PL algorithm was lower than the REQUEST, the convergence rate of the pitch angle calculated by the proposed OP-PL algorithm was half that calculated by the REQUEST algorithm.

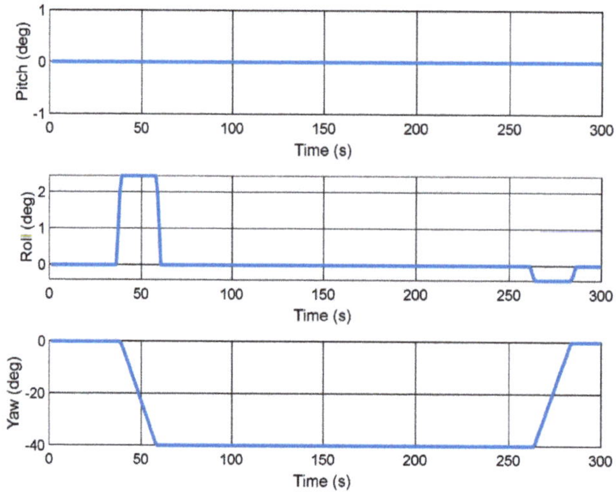

Figure 2. Reference angles.

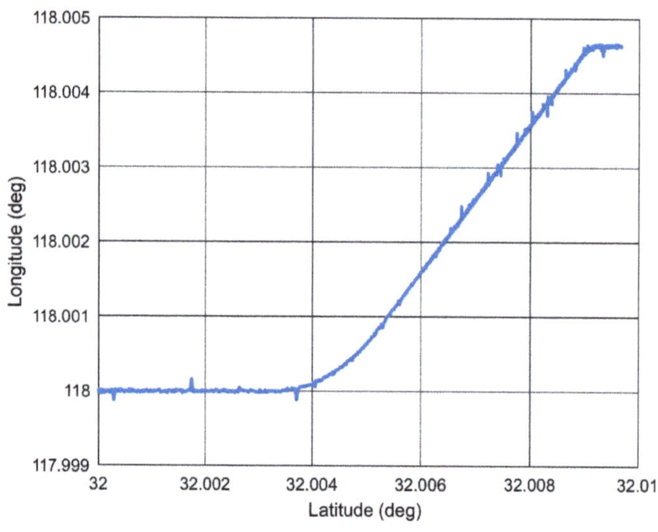

Figure 3. Reference trajectory.

Table 2. Parameter setting of inertial measurement units.

Sensors	Constant Bias	Random Noise
Gyroscopes	0.01 deg/h	0.005 deg/$\sqrt{h}$
Accelerometer	100 μg	100 μg/$\sqrt{Hz}$

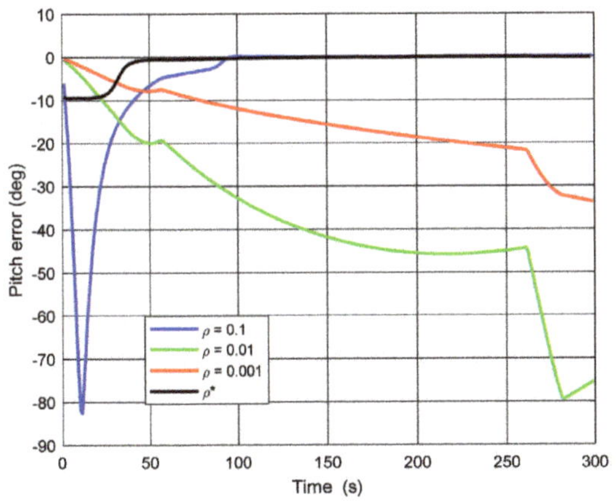

Figure 4. Pitch comparisons of different ρ.

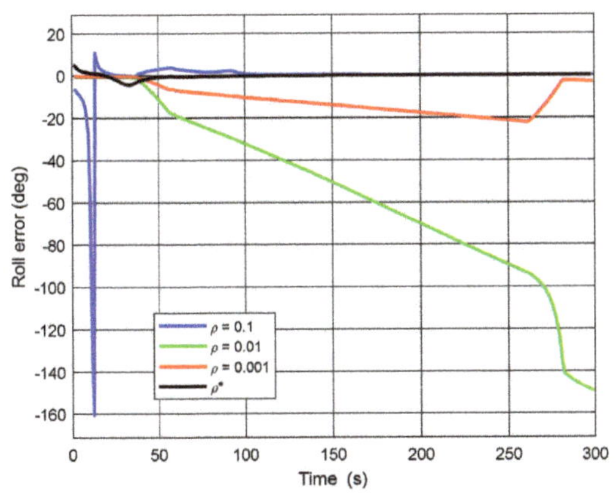

Figure 5. Roll comparisons of different ρ.

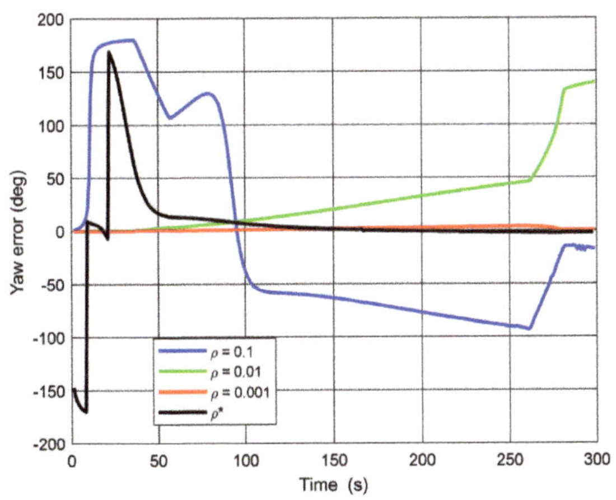

Figure 6. Yaw comparisons of different ρ.

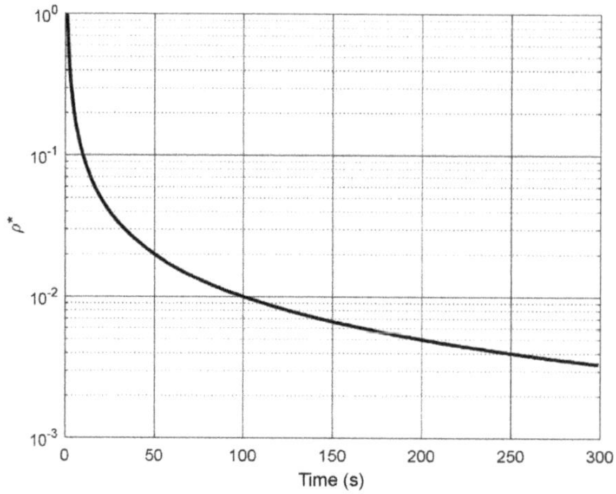

Figure 7. The curve of optimal ρ^*.

Moreover, the roll MEAN errors of the blue and black curves were 0.1875° and 0.1159°, respectively. The roll RMSE errors of the blue and black curves were 0.2114° and 0.1210°, respectively. From the RMSE data, it is evident that the proposed OP-PL algorithm performed better than the REQUEST algorithm for the calculation of the roll angle. It can also be seen from Figure 5 that the proposed OP-PL algorithm was very stable and that the convergence speed was still very fast.

However, when $\rho = 0.1$, the performance of the yaw angle was very poor. Although this value can produce excellent results in calculating the roll angle and the pitch angle, it was completely wrong for calculating the yaw angle. This indicates that the fixed gain was unreliable during the coarse alignment. When the value of ρ was selected as 0.1, the new K-matrix (δK_{k+1}) accounted for a relatively large proportion during the update of the K-matrix. This made it impossible to calculate the yaw angle correctly. If the value of ρ was decreased to 0.001, more error effects were accumulated because the old K-matrix ($K_{k+1/k}$) accounted for a relatively large proportion during the update of the K-matrix. This made it difficult to calculate the pitch and roll angles correctly.

To overcome the drawbacks of the REQUEST algorithm, the proposed algorithm uses a cost function to adjust the fading memory value adaptively. From Figure 7, we can see that, the value is selected as one at the beginning of the filter. Thus, more parts of the new measurements can be used during the update stage to speed up convergence. With increase in alignment time, a relatively low weight is assigned to the new measurement to ensure that the alignment process is steady.

To further verify the superiority of the proposed OP-PL algorithm, we chose a third algorithm for comparison. Based on the obtained simulation results, we chose $\rho = 0.001$ as the fading memory factor in the REQUEST algorithm. Figures 8–10 show the calculation results of the REQUEST, QUEST and the proposed OP-PL algorithm, which are represented by the yellow, green and brown curves, respectively. To better compare the errors among the three algorithms, we list the mean values and root mean square errors of the angle error values in Table 3, and present a box plot which is shown in Figure 11.

In Figures 8 and 9, the performance of the REQUEST algorithm can be seen to be still poor compared with the QUEST and the proposed OP-PL algorithm. The reason for this situation is that the REQUEST is a suboptimal algorithm, and it is hard to find an appropriate value to keep the algorithm stable. Figure 11 confirms the poor performance of REQUEST, where the three angle errors were obviously larger than for the other two algorithms. The MEANs and RMSEs of the pitch and roll angles, which were calculated by the REQUEST algorithm, were also the worst among the three algorithms.

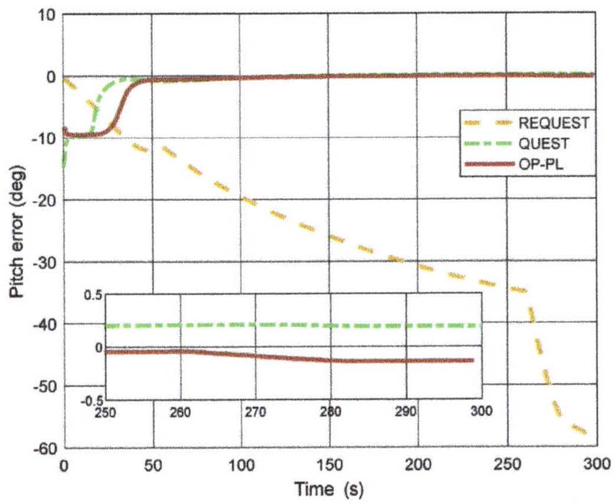

Figure 8. Curves of pitch angle errors.

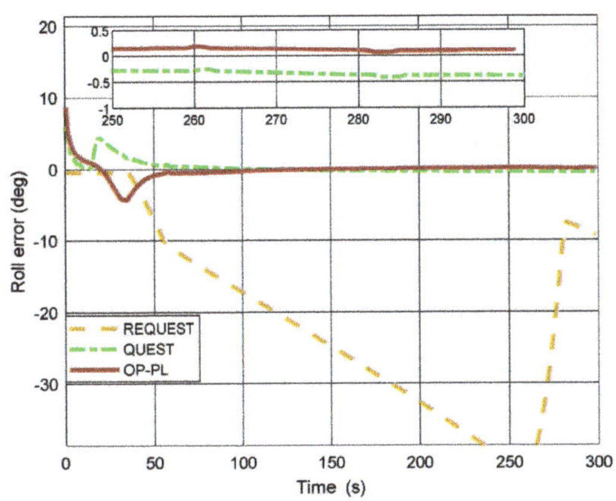

Figure 9. Curves of roll angle errors.

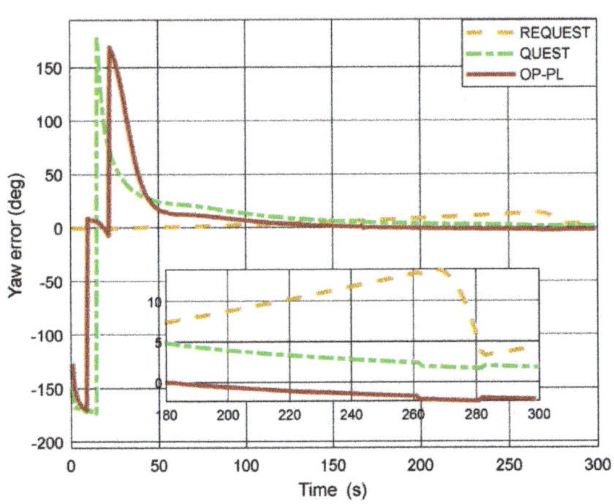

Figure 10. Curves of yaw angle errors.

Table 3. Error statistics of three algorithms in simulation test.

Items			REQUEST	QUEST	OP-PL
Pitch (deg)	1~150 s	MEAN	−14.7921	−1.7156	−2.3169
		RMSE	16.4321	3.5995	4.1962
	150~300 s	MEAN	−36.1971	0.1662	−0.0760
		RMSE	37.5654	0.1715	0.0824
Roll (deg)	1~150 s	MEAN	−11.6315	1.8145	−0.3420
		RMSE	14.3065	12.6798	1.5843
	150~300 s	MEAN	−29.8454	−0.2526	0.1159
		RMSE	31.4920	0.2619	0.1210
Yaw (deg)	1~150 s	MEAN	2.0683	5.7750	10.8685
		RMSE	2.6878	63.6506	57.2915
	150~300 s	MEAN	8.9417	3.5183	−0.9181
		RMSE	9.4477	3.7871	1.4000

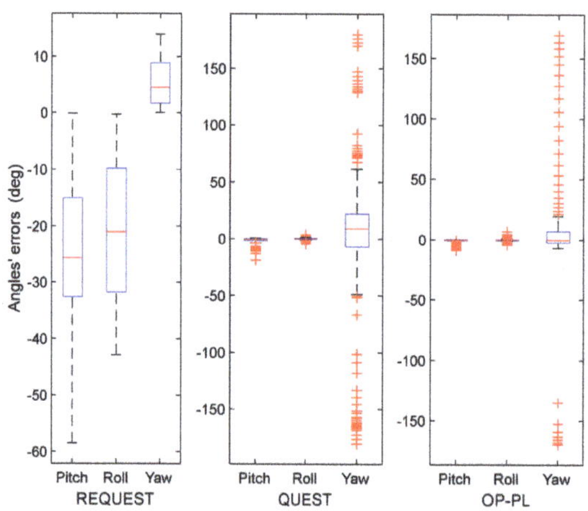

Figure 11. Box plot of angle errors.

Moreover, the OP-PL algorithm retains the merits of the REQUEST algorithm, as it is a time-varying recursive attitude quaternion estimator which preserves the quaternion unit-norm property. Compared with the QUEST algorithm, the OP-PL algorithm has an enhanced ability to reduce the impact of the measurement noise contained in the measurement vectors.

From Figures 8–10, we can see that, although the convergence speed of the OP-PL algorithm was the same as that of the QUEST algorithm, the proposed algorithm had higher accuracy and the accuracy of the three angles were all improved. Moreover, as can be seen from Figure 11, although the outliers of the yaw angle calculated by OP-PL were similar to that calculated by the QUEST algorithm, the proposed algorithm was relatively stable, based on the calculation results from the three angles. The main cause of the outliers in the yaw angle was unstable performance from 0 to 50 s. After the algorithm became stable, the advantages of the proposed algorithm were still evident. The pitch angle, roll angle and yaw angle calculated by the OP-PL algorithm were reduced by approximately 51.95%, 53.80% and 63.03%, respectively, compared with the QUEST algorithm.

Therefore, the above analyses confirm that the OP-PL algorithm had better alignment accuracy than the comparison algorithms.

6. Conclusions and Future Work

This paper proposes an improved in-motion coarse alignment method based on position loci and an Optimal-REQUEST algorithm. The proposed algorithm only uses position information to construct the observation vector, which greatly expands the application scope of the coarse alignment algorithm because all GPS receivers can output the position information. Moreover, the effects of random noise contained in the vector observation are suppressed by the Optimal-REQUEST algorithm which can choose the filter gain adaptively. The results of simulation tests showed that the proposed OP-PL algorithm converged faster and had better accuracy than the comparison algorithms. The pitch angle, roll angle and yaw angle calculated by the OP-PL algorithm were reduced by approximately 51.95%, 53.80% and 63.03%, respectively. However, our work still has some limitations and areas that can be improved. In future work, we plan to verify our proposed algorithm in real vehicle experiments and to extend the application scope of the algorithm to underwater environments.

Author Contributions: Conceptualization, J.W.; Supervision, H.H.; Writing—original draft, J.W.; Writing—review & editing, H.H. All authors have read and agreed to the published version of the manuscript.

Funding: This work is supported by the Natural Science Foundation of Jiangsu Province (BK20221500), the National Natural Science Foundation of China (61703098), and the Fujian Provincial Key Laboratory of Coast and Island Management Technology Study (FJCIMTS2019-03). This work is also supported in part by London Tech Bridge, the APEX Undersea Challenge, the European Regional Development Fund—Industrial Intensive Innovation Programme, the Research Fund from the Science and Technology on Underwater Vehicle Technology Laboratory (2021JCJQ-SYSJJ-LB06905), and the Water Science and Technology Project of Jiangsu Province under grant (2021072, 2021063).

Institutional Review Board Statement: Not applicable.

Informed Consent Statement: Not applicable.

Data Availability Statement: Not applicable.

Conflicts of Interest: The authors declare no conflict of interest.

Appendix A

The definition of P_{xx} is shown below:

$$\mathrm{P}_{xx} = E\left[xx^{\mathrm{T}}\right] \tag{A1}$$

Taking a two-dimensional matrix as an example, the specific calculation method of P_{xx} is:

$$\mathrm{P}_{xx} = \begin{bmatrix} E\left[x_{11}^2\right] + E\left[x_{12}^2\right] & E[x_{11}x_{12}] + E[x_{12}x_{22}] \\ E[x_{11}x_{12}] + E[x_{12}x_{22}] & E\left[x_{12}^2\right] + E\left[x_{22}^2\right] \end{bmatrix} \tag{A2}$$

It is obvious that P_{xx} is a symmetric matrix with positive semi-definite. Considering the vector $\mathrm{X} = \begin{bmatrix} x_{11} & x_{12} & x_{21} & x_{22} \end{bmatrix}^{\mathrm{T}}$ with zero-mean, its covariance matrix is calculated as follows:

$$\mathrm{cov}(\mathrm{X}) \triangleq E\left[\mathrm{XX}^{\mathrm{T}}\right] \tag{A3}$$

Then, we find that the trace of P_{xx} is equal to the trace of $\mathrm{cov}(\mathrm{X})$, namely, $tr(\mathrm{P}_{xx}) = tr(\mathrm{cov}(\mathrm{X}))$. This definition of the P_{xx} matrix expresses the uncertainty of the matrix in a more compact and convenient form. Therefore, we use this definition to evaluate the uncertainty of V_k.

References

1. Titterton, D.; Weston, J.L. *Strapdown Inertial Navigation Technology*, 2nd ed.; IET: London, UK, 2004.
2. Huang, H.; Tang, J.; Liu, C.; Zhang, B.; Wang, B. Variational Bayesian-Based Filter for Inaccurate Input in Underwater Navigation. *IEEE Trans. Veh. Technol.* **2021**, *70*, 8441–8452. [CrossRef]
3. Wang, D.; Xu, X.; Yang, Y.; Zhang, T. A Quasi-Newton Quaternions Calibration Method for DVL Error Aided GNSS. *IEEE Trans. Veh. Technol.* **2021**, *70*, 2465–2477. [CrossRef]
4. Abosekeen, A.; Iqbal, U.; Noureldin, A.; Korenberg, M.J. A Novel Multi-Level Integrated Navigation System for Challenging GNSS Environments. *IEEE Trans. Intell. Transp. Syst.* **2021**, *22*, 4838–4852. [CrossRef]
5. Huang, H.; Tang, J.; Song, R.; Tang, X. A novel matrix block algorithm based on cubature transformation fusing variational Bayesian scheme for position estimation applied to MEMS navigation system. *Mech. Syst. Signal Process.* **2022**, *166*, 108486. [CrossRef]
6. Liu, Y.; Fan, X.; Lv, C.; Wu, J.; Li, L.; Ding, D. An innovative information fusion method with adaptive Kalman filter for integrated INS/GPS navigation of autonomous vehicles. *Mech. Syst. Signal Process.* **2018**, *100*, 605–616. [CrossRef]
7. Wang, J.W.; Chen, X.Y.; Yang, P. Adaptive H-infinite Kalman Filter based on Multiple Fading Factors and its Application in Unmanned Underwater Vehicle. *ISA Trans.* **2021**, *108*, 295–304. [CrossRef]
8. Lin, Y.; Miao, L.; Zhou, Z. An Improved MCMC-Based Particle Filter for GPS-Aided SINS In-Motion Initial Alignment. *IEEE Trans. Instrum. Meas.* **2020**, *69*, 7895–7905. [CrossRef]
9. Xu, J.N.; He, H.Y.; Qin, F.J. A Novel Autonomous Initial Alignment Method for Strapdown Inertial Navigation System. *IEEE Trans. Instrum. Meas.* **2017**, *66*, 2274–2282. [CrossRef]
10. Chang, L.; Li, Y.; Xue, B. Initial Alignment for a Doppler Velocity Log-Aided Strapdown Inertial Navigation System with Limited Information. *IEEE/ASME Trans. Mechatron.* **2017**, *22*, 329–338. [CrossRef]
11. Huang, H.; Wei, J.; Jin, C.; Tang, J. An Improved Initial Alignment Method using Kalman Filtering of the Vectorized K-matrix. In Proceedings of the 2021 International Conference on Sensing, Measurement & Data Analytics in the Era of Artificial Intelligence (ICSMD), Nanjing, China, 21–23 October 2021; pp. 1–5.
12. Zhang, T.; Zhu, Y.; Xu, X.; Wang, J.; Li, Y. In-Motion Coarse Alignment Based on the Vector Observation for SINS. *IEEE Trans. Instrum. Meas.* **2019**, *68*, 3740–3750. [CrossRef]
13. Luo, L.; Huang, Y.; Zhang, Z.; Zhang, Y. A New Kalman Filter-Based In-Motion Initial Alignment Method for DVL-Aided Low-Cost SINS. *IEEE Trans. Veh. Technol.* **2021**, *70*, 331–343. [CrossRef]
14. Xu, X.; Sun, Y.; Gui, J.; Yao, Y.; Zhang, T. A Fast Robust In-Motion Alignment Method for SINS with DVL Aided. *IEEE Trans. Veh. Technol.* **2020**, *69*, 3816–3827. [CrossRef]
15. Liu, X.X.; Liu, X.J.; Song, Q. A Novel Self-Alignment Method For SINS Based on Three Vectors of Gravitational Apparent Motion in Inertial Frame. *Measurement* **2015**, *62*, 47–62. [CrossRef]
16. Liu, Y.; Xu, X.; Liu, X.; Yao, Y.; Wu, L.; Sun, J. A Self-Alignment Algorithm for SINS Based on Gravitational Apparent Motion and Sensor Data Denoising. *Sensors* **2015**, *15*, 9827–9853. [CrossRef]
17. Wang, J.; Zhang, T.; Tong, J.; Li, Y. A Fast SINS Self-Alignment Method Under Geographic Latitude Uncertainty. *IEEE Sens. J.* **2020**, *20*, 2885–2894. [CrossRef]
18. Wu, M.P.; Wu, Y.X.; Hu, X.P. Optimization-based alignment for inertial navigation systems: Theory and algorithm. *Aerosp. Sci. Technol.* **2011**, *15*, 1–17. [CrossRef]
19. Wu, Y.; Pan, X. Velocity/Position Integration Formula Part I: Application to In-Flight Coarse Alignment. *IEEE Trans. Aerosp. Electron. Syst.* **2013**, *49*, 1006–1023. [CrossRef]
20. Xu, X.; Sun, Y.; Yao, Y.; Zhang, T. A Robust In-Motion Optimization-Based Alignment for SINS/GPS Integration. *IEEE Trans. Intell. Transp. Syst.* **2022**, *23*, 4362–4372. [CrossRef]
21. Xu, X.; Xu, D.; Zhang, T.; Zhao, H. In-Motion Coarse Alignment Method for SINS/GPS Using Position Loci. *IEEE Sens. J.* **2019**, *19*, 3930–3938. [CrossRef]
22. Yao, Y.; Xu, X.; Zhang, T.; Hu, G. An Improved Initial Alignment Method for SINS/GPS Integration with Vectors Subtraction. *IEEE Sens. J.* **2021**, *21*, 18256–18262. [CrossRef]
23. Oshman, Y.; Bar-Itzhack, I.; Choukroun, D. Optimal-REQUEST algorithm for attitude determination. *J. Guid. Control Dyn.* **2004**, *27*, 418–425.

applied sciences

MDPI

Article

The LOS/NLOS Classification Method Based on Deep Learning for the UWB Localization System in Coal Mines

Yuxuan Zhao † and Manyi Wang *,†

The School of Mechanical Engineering, NanJing University of Science and Technology, Nanjing 210094, China; zhaoyx@njust.edu.cn
* Correspondence: manyi.wang@njust.edu.cn
† These authors contributed equally to this work.

Abstract: A localization system is one of the basic requirements for coal mines. Ultra-wideband (UWB), as a technology with broad application prospects, is considered to have great potential in the absence of satellite signals, especially in the underground mine environment, as it can provide rescue assistance. However, state-of-the-art UWB position systems in coal mines cannot efficiently differ the line-of-sight from all communication links, which results in deterioration of the localization accuracy. In this paper, we propose a LOS/NLOS classification method based on a deep learning algorithm. Specifically, we utilize the Generative Adversarial Networks (GAN) to generate diagnostic data for frame transmission under non-line-of-sight (NLOS) condition. Then, a Convolutional Neural Network (CNN) is adopted to identify the NLOS communication. Finally, the trilateral centroid positioning algorithm (TCPA) based on ranging data is used to verify the effect of our method for a localization system in coal mines. Field experiments show that our method can accurately differ the LOS/NLOS with the accuracy of 91.19%. The TCPA algorithm with our method can obtain 3.11% improvement compared with the scenario without using our method.

Keywords: Ultra-Wide-Band; non-line-of-sight identification; generative adversarial network; convolutional neural network; trilateral location

1. Introduction

In recent years, with the continuous development of wireless communication and wireless sensor network, wireless localization technology is widely used in disaster first aid, target tracking, vehicle navigation, pedestrian navigation, and other fields. Outdoor localization technology based on a global navigation satellite system has become viable. However in indoor and underground scenes without satellite signals, accurate localization has become a problem to be solved [1–3]. Ultra-wideband (UWB) has the advantages of high localization accuracy, high communication rate , strong multi-path resolution, and low power consumption. Thus, UWB has a good application prospect in the wireless localization field without satellite signals [4–8].

Facing the environment of an underground mine, in the existing UWB positioning systems, most of the channel impulse response (CIR) data of UWB are selected as the data for classifying NLOS [9–11]. However, because the length of the CIR data is too long, when the tag capacity is large, it may not be possible to obtain the full-length CIR data. In addition, in the actual use of UWB positioning system, the number of LOS data is usually larger than that of NLOS data. Even if all NLOS data is misidentified as LOS data, a high recognition rate can be achieved [12]. Thus, although NLOS data are scarce, they are more important than LOS data. Finally, after NLOS is identified, the distance value measured by UWB still contains other errors, and the positioning accuracy cannot be improved without eliminating them [13–15].

To address the aforementioned issues, we develop a LOS/NLOS classification method based on deep learning for the UWB localization system in coal mines. The problem of imbalanced classification and NLOS classification can be overcome in this study, and a person in a coal mine can be precisely located. However, it is worth mentioning that we merely introduced map limitations after the localization process to fix the results, while the actual situation needs to be considered much more than that. Nevertheless, the methods proposed in this paper, as a foundation, are constructive.

Our contributions in this paper are four-fold:

1. We propose a NLOS recognition method based on deep learning to distinguish whether a UWB signal transmission is blocked in coal mines. We use the diagnostic data in theDW1000 register as the data feature of deep learning, and analyze the classification performance of multilayer perceptrons and a convolutional neural network;
2. We propose a method for data enhancement based on generative adversarial networks (GAN) to solve the problem of class imbalance of data samples. We use this method to generate a new data set and compare the performance of the new data set with the original data set on the same classifier;
3. We propose a trilateral centroid positioning algorithm, and propose how to calculate the location result of the algorithm when a non-ideal situation occurs. Finally, we corrected the results with the map constraints;
4. We designed static and dynamic experiments to verify our system, and evaluated the proposed techniques through experiments and analysis.

The rest of this paper is structured as follows. Section 2 reviews the related works and presents the system architecture that we propose. Sections 3 and 4 describe the mathematical principles of the method used in the system. Section 5 introduces our experimental system and analyzes and discusses the experimental results. Finally, we present our conclusions in Section 6.

2. Related Work and Positioning System Structure

In order to solve the above problems, many people have begun to do research on a UWB positioning system. C Jiang et al. selected the CIR data as input of CNN-LSTM model to classify NLOS/LOS [16]. CL Sang et al. selected 12 features of the UWB signals as data features, and compared the data classification performance of the SVM, RF, and MLP models [17]. By et al. proposed to use a GAN and auto-encoder to enhance the training classification model and achieve better recognition performance [18]. Song Bo et al. proposed a fast imbalanced binary classification method based on moments (MIBC) to identify the NLOS signals [12]. K Kaemarungsi et al. adopted a fingerprint algorithm to do localization [19]. H Van Trees et al. proposed to utilize the prior information to improve the localization accuracy of the system [20]. A Poulose et al. used a deep long short-term memory network to extract more features from user distance information [21]. S Djosic et al. designed a hybrid algorithm based on fingerprint algorithm to identify NLOS in different environments [22].

After referring to the related work of others, we proposed our positioning system. Figure 1 shows the frame diagram of our positioning system.

As shown in Figure 1, our positioning system consists of three components: UWB communication, UWB data processing, and localization algorithm. When a coal mine worker enters the positioning area with a UWB tag, the tag will actively communicate with three nearby anchors. Each anchor calculates the distance value between the anchor and the tag based on the time-of-flight principle, and the register in the anchor device will record the frame transmission diagnostic data of the last data frame received during each communication. After completing each ranging, the anchor will transmit ranging data and frame transmission diagnostic data to the host computer. We trained a convolutional neural network that can utilize frame-transfer diagnostic data for NLOS/LOS classification. Then the linear model and Kalman filter are used to eliminate the errors in the ranging data. When the ranging data of the three anchors are processed through the above steps, the localization result is obtained by using the trilateral centroid localization algorithm. Finally, we correct our results with map constraints and display the results. We will describe the whole components in detail in the next two sections.

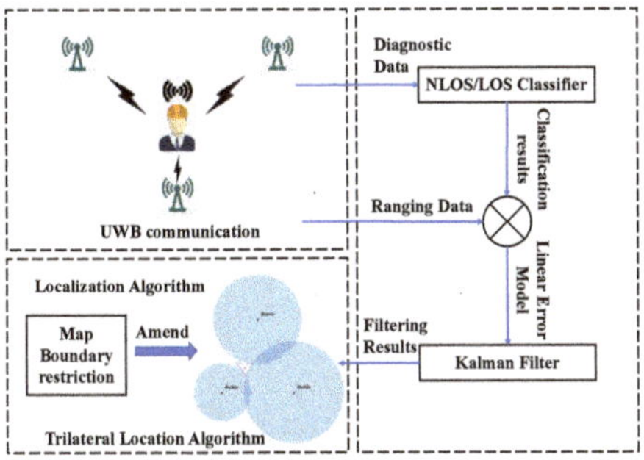

Figure 1. Frame diagram of our positioning system.

3. Nlos and Los Classification Based on Deep Learning Algorithm

3.1. Data Augmentation by Generative Adversarial Networks

For the problem of class imbalance, we adopt the generative adversarial networks (GAN) for data augmentation. A typical theoretical GAN network structure is shown in Figure 2.

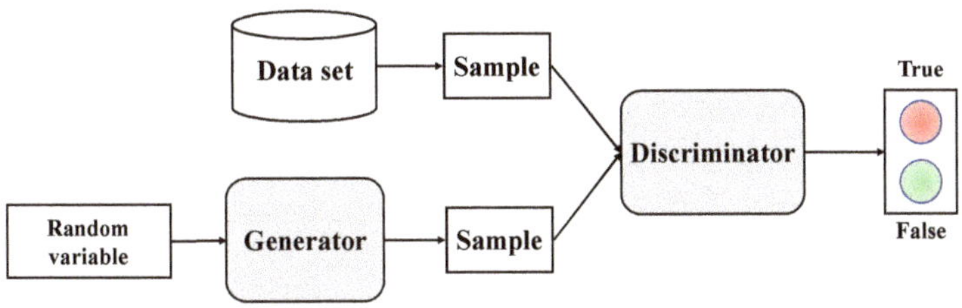

Figure 2. Structure of a generative adversarial network.

GAN consists of two main parts: generator and discriminator. The main function of the generator is to generate the sample data. The main function of the discriminator is to distinguish the receive sample data from the "spurious data" generated by the generator. The generator and the discriminator are trained at the same time, competing against each other. The loss function of GAN (cross entropy loss function) can be expressed as [23]:

$$\min_{G}\max_{D} V(D,G) = \mathbb{E}_{x\sim p_{data}(x)}[\log D(x)] + \mathbb{E}_{x\sim p_z(z)}[\log(1 - D(G(z)))] \quad (1)$$

where $p_{data}(x)$ represents the distribution of real data and $p_z(z)$ represents the distribution of generated "spurious data".

WGAN-GP network is a proposed variant based on the basic GAN network, which can solve the problem of unstable training of the original GAN network. The optimization function of WGAN-GP can be expressed as:

$$L = \underset{\tilde{x}\sim\mathbb{P}_g}{\mathbb{E}}[D(\tilde{x})] - \underset{x\sim\mathbb{P}_r}{\mathbb{E}}[D(x)] + \lambda \underset{\hat{x}\sim\mathbb{P}_{\hat{x}}}{\mathbb{E}}[(\|\nabla_{\hat{x}}D(\hat{x})\|_2 - 1)^2] \quad (2)$$

where $\tilde{x} \longleftarrow G(z), \hat{x} \longleftarrow \varepsilon x + (1-\tilde{x}), \varepsilon \sim U[0,1], \mathbb{P}.$ represent the distribution of $\cdot$.

WGAN-GP adopts a new loss function, and the difference before the plus sign is called the Wasserstein distance. It is used to measure the distance or effort required to transform one distribution into another. The formula after the plus sign is called the Gradient penalty, which is a penalty term for the weight of D.

In GAN networks, the loss function for generators must be reduced, whereas the maximum loss function for discriminators must be optimized. By the max-min game, the generator and discriminator are trained alternately in a loop until the Nash equilibrium point is reached. At this time, after the Gaussian white noise is transformed by the generator, it can be used as a part of the data set for subsequent work to achieve the effect of data augmentation.

In order to solve the problem of class-imbalance, we train a GAN. We built the generator model and the discriminator model, shown in Figure 3. The coefficient of the penalty term λ is taken as 0.3. During the training, the generator is responsible for generating "spurious data" by randomly generating latent data of shape [*batch_size*, 4]. The discriminator receives both "spurious data" generated from generator and *batch_size* samples from the NLOS data.

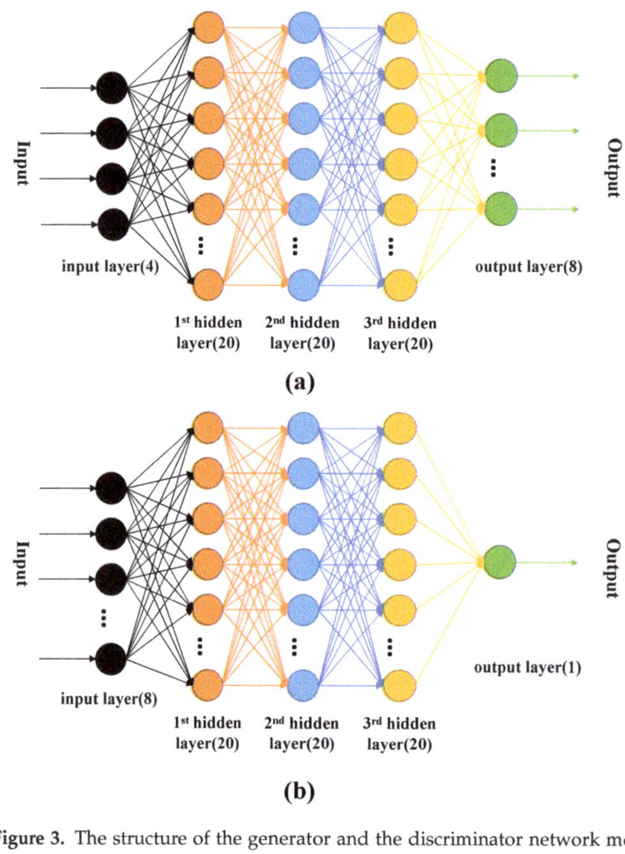

Figure 3. The structure of the generator and the discriminator network models. The number in parentheses after the layer name is the number of neurons contained in the layer. (**a**) Structure of the generator, (**b**) Structure of the discriminator.

3.2. *Multilayer Perceptrons for Classification Problems*

Neural networks inspired by the human brain, also known as connectivity models, are not a recent technology. In 1958, Frank Rosenblatt proposed the first neural network with learning ability, called perceptron [24].

Artificial neurons are at the heart of all neural networks. Figure 4 shows a schematic of the basic structure of a neuron. It consists of two main parts: an adder that weights the sum of all inputs to the neuron, and a processing unit that produces an output according to a predefined function called the activation function. Each neuron has its own set of weights and thresholds (bias), which it learns through different learning algorithms.

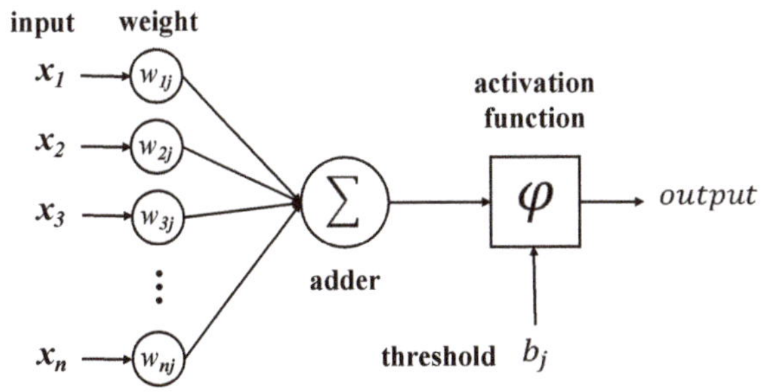

Figure 4. The basic structure of a neuron.

When only one of these neurons exists, it's called a perceptron. When there are many such layers, it is called a multilayer perceptron (MLP). Figure 5 shows a schematic of the MLP structure. MLP have one or more hidden layers. These hidden layers have different numbers of hidden neurons. Each neuron in the hidden layer has the same activation function.

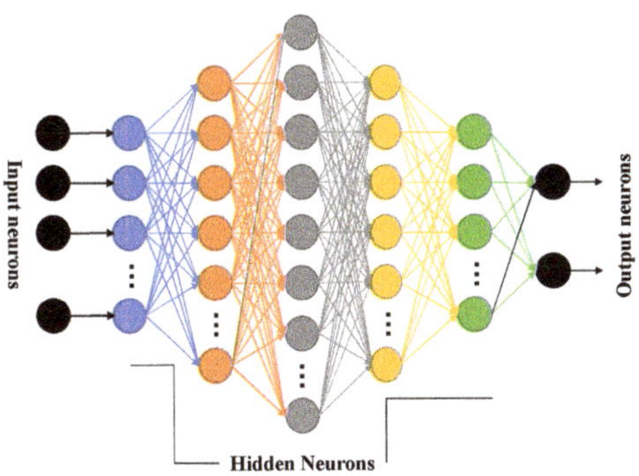

Figure 5. Schematic of the MLP structure.

MLP learn by supervised learning algorithms, that is, they provide the network with the ideal output of training data sets. At the output end, an error function or objective function $J(W)$ is defined so that the objective function will be minimal after the network has fully learned all the training data. A neuron can solve a linearly separable problem, while a learned MLP can solve a linearly indivisible problem.

3.3. Convolutional Neural Networks for Classification Problems

The convolutional neural network (CNN) is a kind of feedforward neural network with deep structure and convolution computation. A CNN has the ability of representation learning, which can carry out shift-invariant classification of input information according to its hierarchical structure, and can carry out supervised and unsupervised learning. Figure 6 shows the basic structure of CNNs.

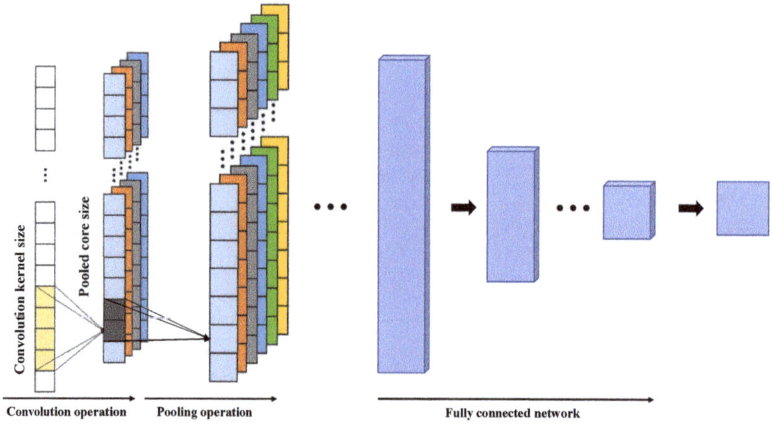

Figure 6. Structure of convolutional neural networks.

The convolutional layer is the most important layer in the whole neural network. The most core part of this layer is the convolutional kernel. The convolution process is designed to learn more complex features hidden in the data. After each convolution layer, a nonlinear layer (or activation layer) is usually applied immediately. The activation layer is generally followed by the pooling layer, which may substantially minimize the size of the matrix. Finally, there is a full connection layer before the output layer, which can integrate and normalize highly abstract features to learn these nonlinear combination features.

It is assumed that the CNN structure is composed of L hidden layers, where the input layer is denoted as 0-th layer, the output layer is $(L+1)$-st layer, and the output layer has K nodes, which represent K possible classes. The relation between the input of each layer and its output can be given as:

$$x^{(l+1)} = g(W^{(l)}x^{(l)}) \tag{3}$$

where $g(\cdot)$ is the convolution layer, pooling layer, or full connection layer operation, $W^{(l)}$ and $x^{(l)}$ are the weights and the outputs of the l-th layer, respectively. The weights and inputs of hidden layers can be represented as$W_L = (W^{(1)}, W^{(2)}, \cdots, W^{(L)})$ and $x_L = (x^{(1)}, x^{(2)}, \cdots, x^{(L)})$, respectively. Therefore, the input and output relationship of CNN can be written as follows [25]:

$$x^{(L+1)} = \underbrace{W^{(L)}W^{(L-1)}\ldots W^{(1)}}_{W_L}x^{(0)} \tag{4}$$

Figure 7 shows the MLP and the CNN structures we designed, for the MLP model, we choose the *sigmoid* function as the activation function of each neuron, and, for the CNN model, we choose the *relu* function as the activation function of the hidden layer neurons.

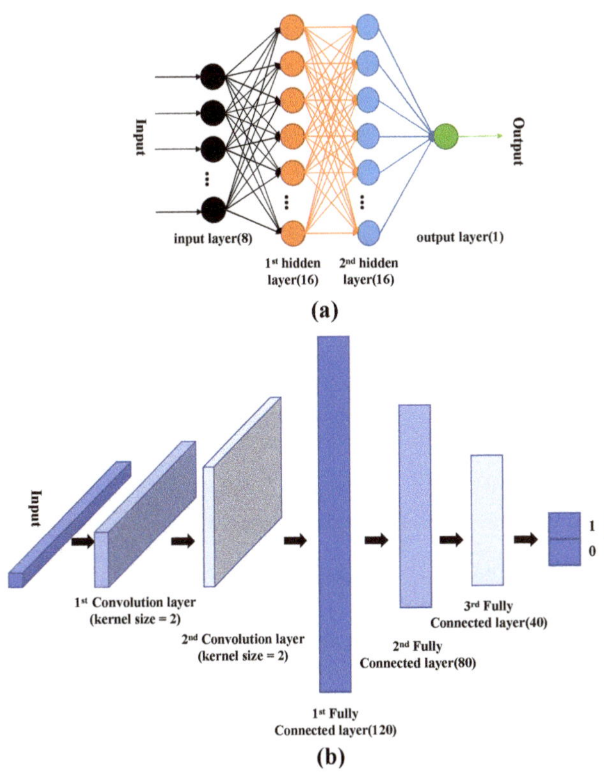

Figure 7. The structure of the MLP and CNN models. The number in parentheses after the layer name is the number of neurons contained in the layer. (**a**) Structure of the MLP, (**b**) structure of the CNN.

4. Location Algorithm Based on Ranging

4.1. Composition of Ranging Errors

Since ranging needs to calculate the signal flight time, it is obvious that the propagation channel will reduce the ranging accuracy. The distance measurement error (DME) can be defined as:

$$\varepsilon_W(d) = \widehat{d_W} - d \tag{5}$$

where $\widehat{d_W}$ is the distance value calculated by the anchor based on DS-TWR, and d is the true distance between TAG and ANCHOR. The DME comes from multipath errors, errors in the process of measuring the real distance between the tag and the anchor, NLOS errors, and device antenna delay errors [26]. The random variable ξ is used to model the occurrence of NLOS conditions, and the random variable takes a value of 1 when the NLOS condition occurs, otherwise the value is 0. Specifically, it can be expressed as:

$$\varepsilon_W(d) = \varepsilon_A + \varepsilon_T + \varepsilon_{M,W}(d) + \xi * \varepsilon_{N,W}(d) \tag{6}$$

where ε_A is the error caused by the UWB device error, including antenna delay, crystal error, etc. Its value fluctuates to some extent. ε_T is the error caused by the operation or reading of the measuring instrument during the actual measurement of d, which approximately obeys the normal distribution, $\varepsilon_{M,W}(d)$ is the error caused by multipath, and $\varepsilon_{N,W}(d)$ is the error caused by NLOS condition. Figure 8 shows the NLOS condition and how the multipath error is generated.

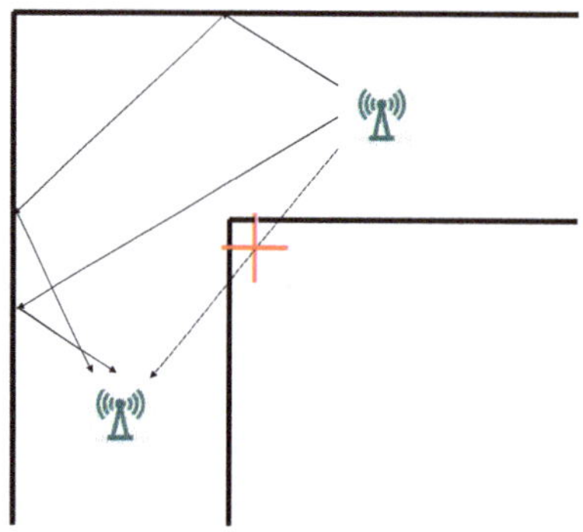

Figure 8. Schematic diagram of NLOS condition and how the multipath error is generated. The black dashed arrow path represents the NLOS condition, and the black solid arrow path represents the multipath transmission.

4.2. Linear Model and Kalman Filter

There are many ways to eliminate DME. The linear model equation used in this paper is as follows:

$$y = (1-\xi)(k_{los} \cdot x + b_{los}) + \xi(k_{nlos} \cdot x + b_{nlos}) \tag{7}$$

where y is the real distance between the tag and the anchor, x is the distance value calculated by the anchor based on TOF, ξ is the state quantity of NLOS, and k_{los}, b_{los}, k_{nlos}, and b_{nlos} are parameters corresponding to LOS and NLOS, respectively. The value of ξ is based on the corresponding line-of-sight conditions during data collection, 1 for NLOS, and 0 for LOS.

Kalman filter is an algorithm that uses a series of data observed over time, which contains noise and other inaccuracies, to estimate unknown variables with more accuracy [27]. For ranging, the positioning system is a linear system. A discrete Kalman filter can be used to filter the ranging value. According to Equation (6), there are multiple errors in the DME. ε_A error includes clock drift error, and ε_T approximately obey normal distribution. Kalman filter can eliminate the errors caused by both and makes the data more stable.

We define the system state variable as $X_k \in R^n$, the system control input as U_k, and the system process excitation noise as W_k.Then the state equation of the system is:

$$X_k = AX_{k-1} + BU_k + W_k \tag{8}$$

The observed variable is $Z_k \in R^m$, and the observed noise is V_k. Thus, the measurement equation can be obtained as follows:

$$Z_k = HX_k + V_k \tag{9}$$

where H is the observation model matrix, which maps the real state space to the observation space. Assuming that W_k and V_k are mutually independent and normally distributed white noise, the process excitation noise covariance matrix is Q, and the observation noise covariance matrix is R, that is:

$$W_k \sim N(0, Q) \tag{10}$$

$$V_k \sim N(0, R) \tag{11}$$

The discrete Kalman filter includes the prediction process and update process. Prediction equations can be expressed as follows:

$$\widehat{X}_k^- = A\widehat{X}_{k-1} + B\widehat{U}_{k-1} \tag{12}$$

$$P_k^- = AP_{k-1}A^T + Q \tag{13}$$

Update equations are expressed as follows:

$$K_k = P_k^- H^T (HP_k^- H^T + R)^{-1} \tag{14}$$

$$\widehat{X}_{k-1} = \widehat{X}_k^- + K_k(Z_k - H\widehat{X}_k^-) \tag{15}$$

$$P_k = (I - K_k H)P_k^- \tag{16}$$

where K_k, $\widehat{X}_k$, P_k, I is the Kalman gain matrix, the optimum filter value, the filter deviation matrix, and the unit matrix, respectively.

The prediction process mainly uses the prediction equation to establish a priori estimation of the current state, and predicts the current state variable and the value of the error covariance estimate forward in time, so as to construct a priori estimate for the next time state; the more trust process is responsible for feedback, and using the prediction equation establishes an improved posterior estimate of the current state based on a priori estimates of the prediction process and current measured variables.

4.3. Trilateral Centroid Positioning Algorithm

The positioning method of the UWB positioning system can use the trilateral centroid positioning algorithm (TCPA). Figure 9 shows the schematic diagram of the original trilateral centroid positioning algorithm.

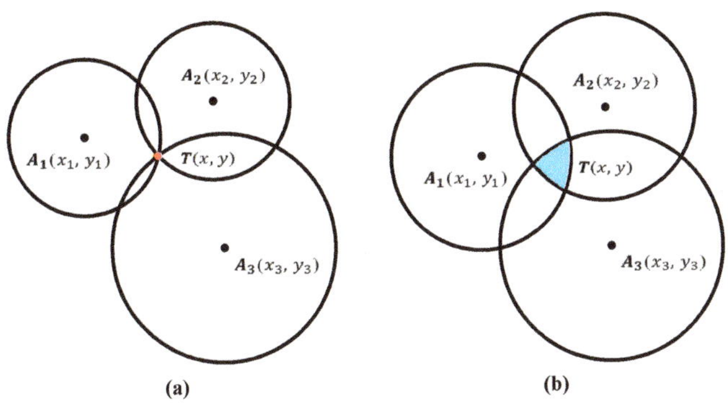

Figure 9. Schematic diagram of the trilateral centroid positioning algorithm. The black point represents the position of the anchor. (**a**) Ideal case: Three circles intersect at a point. (**b**) Non-ideal case: Three circles do not intersect at a point. The blue area indicates an error in the distance measurement.

In the two-dimensional plane coordinate system, it is known that the coordinates of the three anchors are $A(x_1, y_1), A(x_2, y_2), A(x_3, y_3)$, and the coordinates of the tag are $T(x, y)$. The distances between the tag and the three anchors are r_1, r_2 and r_3, respectively. Taking the three anchor coordinates as the center and the corresponding r as the radius, three circles can be formed. They have the following relationship:

$$\begin{cases} (x - x_1)^2 + (y - y_1)^2 = r_1^2 \\ (x - x_2)^2 + (y - y_2)^2 = r_2^2 \\ (x - x_3)^2 + (y - y_3)^2 = r_3^2 \end{cases} \tag{17}$$

Ideally, as shown in the Figure 9a, three circles intersect at a point. However, when there are errors in r_1, r_2, and r_3 as shown in the Figure 9b, they will not intersect at one point. Tag is in the blue area, where the optimal position needs to be found.

The principle of TCPA is to use the centroid of a triangle determined by pairs of circles as the location result. There are five types of relations between two circles: intersection, inner tangent, outer tangent, inner separation, and the outer separation. Figure 10 shows the schematic diagram of the two positioning processes including these five relations.

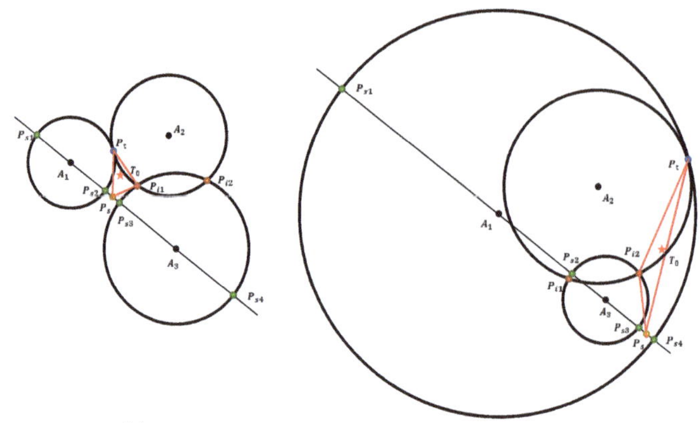

Figure 10. The special case in the TCPA where the three circles do not intersect at a point. The triangle formed by the red line segment is the triangle selected by the trilateral centroid positioning algorithm, and the red five-pointed star is the centroid of the triangle, which is also the positioning result. (**a**) Circle A1 and circle A3 are outer separated. (**b**) Circle A1 and circle A3 are inner separated.

When two circles are tangent, the tangent point(P_t in the Figure 10a,b) is directly used as the vertex determined by the two circles.

When two circles are separated, the vertex determined by the two circles is the midpoint at which they are closest to each other, just as point P_s in Figure 10a,b.

When the two circles intersect, the vertex determined by the two circles is the intersection point at which the sum of the sides of the triangle with the other vertices determined or to be determined is the smallest, just as the point P_{i1} in Figure 10a and the point P_{i2} in Figure 10b.

After determining the vertex of the triangle, we calculate the centroid of the triangle, which is the positioning result of TCPA. When we calculate the centroid coordinates of a triangle determined by the TCPA, it may fall outside the boundary of the map if we display it directly. So it needs to be corrected when it falls outside the boundaries of the map. Figure 11 shows a schematic diagram of the corrected positioning results.

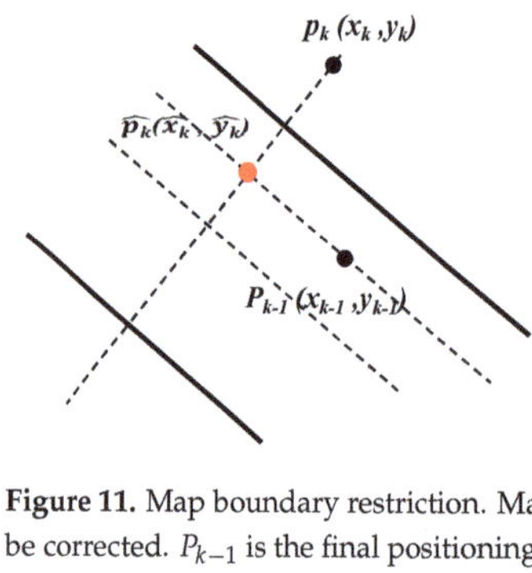

Figure 11. Map boundary restriction. Map restriction scheme for the positioning results that need to be corrected. P_{k-1} is the final positioning result at $k-1$, P_k is the centroid coordinates of the triangle determined by the trilateral centroid positioning algorithm at k, and $\widehat{p}_k$ is the corrected positioning result.

We oass point P_k to take a normal line of the map boundary and extend it. The corrected positioning result should fall on the normal line; pass point P_{k-1} to take a parallel line (or curve) of the map boundary, then it will intersect the normal at point $\widehat{p}_k$. Point $\widehat{p}_k$ must fall within the map boundaries, therefore taking it as the corrected result.

5. Experiment, Results and Discussion

5.1. Experiment Set Up and Data Collection

5.1.1. Hardware Platform

As shown in the Figure 12, we use the device based on DW1000 [28] as the tag and the anchor. Ranging using a TOF based ranging method.

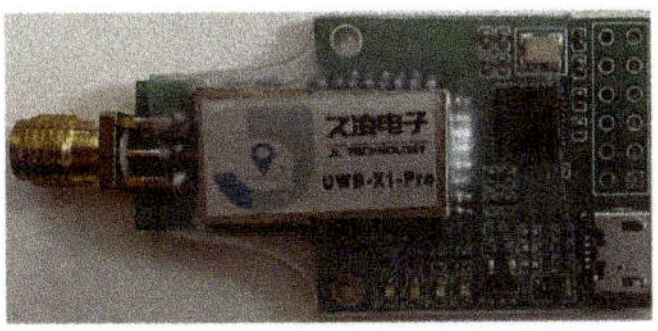

Figure 12. The UWB node. The device we use is produced by China Jiuling electronics manufacturer, whose Chinese name is printed on the surface of the device.

5.1.2. Data for Training Network Model

We use the diagnostic data as the input data for our classifier. So both the MLP model and the CNN model we established have eight input neurons. As shown in Table 1, eight symbols of the transmitting data are involved in our diagnostic data.

Table 1. Symbols of the frame transmit data we used to read from the registers.

Serial Number	Symbols	Meaning
1	firstPathAmp1	the magnitude of the accumulator tap at the index 3 beyond the integer portion of the rising edge FP_INDEX reported in Register
2	firstPathAmp2	the magnitude of the accumulator tap at the index 2 beyond the integer portion of the rising edge FP_INDEX reported in Register
3	firstPathAmp3	the magnitude of the accumulator tap at the index 1 beyond the integer portion of the rising edge FP_INDEX reported in Register
4	stdNoise	the standard deviation of the noise level seen during the LDE algorithm's analysis of the accumulator data.
5	maxGrowthCIR	a growth factor for the accumulator which is related to the receive signal power.
6	firstPathIdex	the position within the accumulator that the LDE algorithm has determined to be the first path.
7	rxPreamCount	the number of symbols of preamble accumulated.
8	C	the Channel Impulse Response Power value reported in the CIR_PWR field of Register file

According to the DW1000 Software API Guide and the DW1000 user manual, in the above diagnostic data, *stdNoise* reported in the register file may be used to give a measure of the noise associated with this and the received frame's timestamp measurement. firstPathAmp1, firstPathAmp2, firstPathAmp3 are 16-bit unsigned values that are part of reporting the magnitude of the leading edge signal seen in the accumulator data memory during the LDE algorithm's analysis. In addition to stdNoise, other data types will be used to calculate the first path power level and the receive power level. High noise may mean that the real first path is irretrievably buried in the noise. Comparing the noise with the First Path Amplitude can give additional indication as to the quality of the first path measurement. So those features can be used in assessing the quality of the received signal. In order to obtain the LOS sample data and the NLOS sample data with these eight characteristics, we designed an experimental system to collect the LOS sample data and the NLOS sample data, respectively. Figure 13 shows a top view of our experimental scenario.

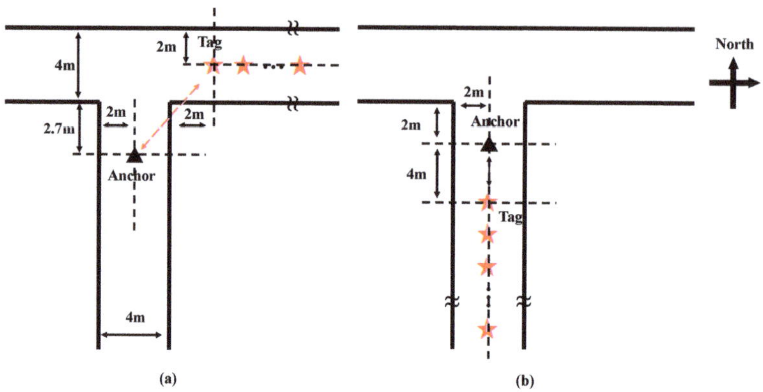

Figure 13. An experimental scenario for collecting data for training the model network. The black thick solid line indicates the boundary of the environment, and the red bidirectional arrows indicate the mutual ranging devices (dotted line indicates the NLOS, solid line indicates the LOS). The red five-pointed star indicates the position of the tag. (**a**) Data collection for non-line-of-sight situations, (**b**) Data collection for line-of-sight situations.

We marked the location of the tag, before the data is collected. For the collection of NLOS data, we moved 1 m to the east of the map each time, and collected a group of several NLOS samples at each location. A total of 11 moves can collect 12 sets of data. For the collection of LOS data, we move 2 m to the south of the map each time, and collect a set of LOS sample data of a certain number at each position, moving a total of 53 times, that is, 54 sets of data can be collected. Each sample data includes the ranging values and eight features described in the previous paragraph. Table 2 shows the number of sample data collected by out experiments.

Table 2. The amount of sample data needed to train the model network.

NLOS/LOS	Number of Set	Total
NLOS	12	1812
LOS	54	6967
total	-	8779

5.1.3. Data for Testing Position System

To test our designed localization system, we designed static and dynamic experiments to examine our system, respectively. Figure 14 shows the concept diagrams for our static and dynamic experiments.

For the static experiments, we fixed the three anchors in the positions shown in Figure 14a, The tag is placed in one of the three positions (red position, blue position, green position) at a time. The bidirectional arrows in the same color as the tag indicate the anchor ranging from the tag (dotted line indicates the NLOS, solid line indicates the LOS). When the tag is in each position, we collect the data transmitted by the three anchors. Each sampling data includes ranging value and eight features, and the laptop records the sample data from each anchor separately.

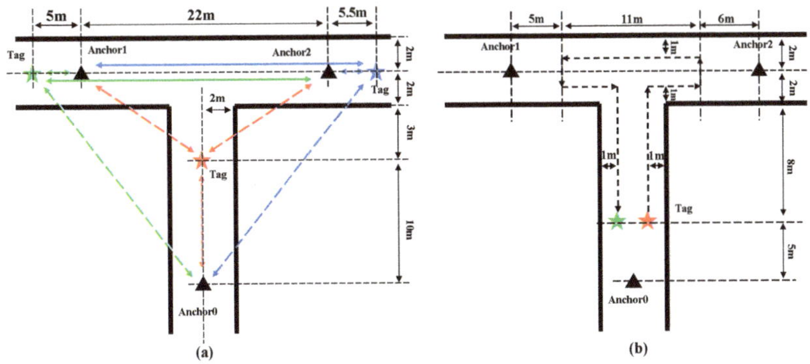

Figure 14. (**a**,**b**) are the concept diagrams of the static experiment and dynamic experiments, respectively. The black thick solid line indicates the boundary of the environment, and the five-pointed star indicates the position of the tag at different positions.

As shown in Figure 14b, for the dynamic experiment, we have the same setup as in the static experiment, where a person carries a tag from the red five-pointed star position as far as possible along the path of the dotted black arrow at a uniform speed until the green pentagram position ends. During the movement process, each anchor transmits the sampling data containing the ranging value and eight features of each ranging to the host computer.

Table 3 shows the amount of the data we collected from the static and the dynamic experiments.

Table 3. The amount of data collected by the experiment to test our designed positioning system. The data is obtained by one tag and three anchors, each performing a ranging as a group.

Static/Dynamic Experiment	Position	Amount of Sample Data (Group)
Static experiment	red	70
	blue	68
	green	68
Dynamic experiment	-	104

5.2. Experimental Results and Discussion

5.2.1. Network Model Training Results and Discussion

The NLOS data and the LOS data are 1812 and 6967, respectively. There is five times of discriminator training in one training, and one training of the generator, with a total of 1000 times of training. According to the theory of GAN, we want to maximize the loss function of the discriminator as much as possible. The loss function of the discriminator is shown in Figure 15. It can be seen that the loss function of the discriminator gradually increased and stabilized near the maximum value.

After the GAN is trained, we put 2000 random Gaussian noise data into the generator to generate 2000 samples. We copied the original dataset and added those 2000 samples to form a new data set. Now our original data set includes 6967 LOS data and 1812 NLOS data, and the new data set includes 6967 LOS data and 3812 NLOS data. The two data sets are divided into the training set and the test set according to the ratio of 7:3. We use these training sets to train the MLP model and CNN model, and test our model with the same test set.

After the network model is trained, on the same test set, the performance of MLP model and CNN model trained by the training set of the original data set is 88.0790% and 90.8125%, respectively. The performance of the MLP model and CNN model trained by the training set of the new data set is 89.1420% and 91.1921% respectively. We further evaluate the performance of our model using the $classification\_report()$ function from the machine learning *scikit-learn* library. The performance of our trained MLP models and the performance of our trained CNN models are shown in the Tables 4 and 5, separately.

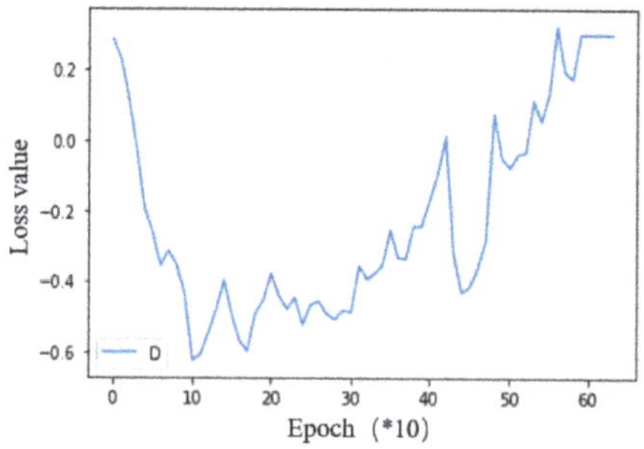

Figure 15. Variation of the discriminator loss function during the training.

Table 4. Performance of our trained MLP models.

Using the Data Generated by the Generator		Precision	Recall	f1-Score	Support
NO	LOS	0.89	0.96	0.93	2092
	NLOS	0.80	0.56	0.66	542
	accuracy			0.88	2634
YES	LOS	0.91	0.95	0.93	2092
	NLOS	0.78	0.65	0.71	542
	accuracy			0.89	2634

Table 5. Performance of our trained CNN models.

Using the Data Generated by the Generator		Precision	Recall	f1-Score	Support
NO	LOS	0.93	0.96	0.94	2092
	NLOS	0.82	0.72	0.76	542
	accuracy			0.91	2634
YES	LOS	0.94	0.95	0.94	2092
	NLOS	0.79	0.77	0.78	542
	accuracy			0.91	2634

As can be seen from Tables 4 and 5, the precision of the MLP model trained by the training set of the original and new data sets for the NLOS class is 0.80 and 0.78, the recall is 0.56 and 0.65, and the f1-score is 0.66 and 0.71, respectively. The accuracy of all samples are 0.88 and 0.89, respectively. The precision of the CNN model trained by the training set of the original data set and the new data set for the NLOS class is 0.82 and 0.79, the recall is 0.72 and 0.77, and the f1-score is 0.76 and 0.78, respectively. The accuracy of all samples is 0.91. After the GAN model is used to enhance the data, the recall of NLOS data in MLP model and CNN model is increased by 9% and 5% respectively, and the f1-score is increased by 5% and 2% respectively. This means that GAN data enhancement can improve the performance of the network model. In horizontal comparison, after the GAN model is used to enhance the data, the CNN model performs better than the MLP model in terms of the recall and f1-score of both LOS and NLOS class.

In conclusion, data enhancement by GAN can improve the performance of the network model. By comparing the performance of MLP model and CNN model, it can be seen that the CNN model has better performance.

5.2.2. Localization System Test Results and Discussion

In the static experiment, we collected about 70 data sets at three positions, respectively. Figures 16 and 17 show the results and localization errors of this static experiment. Since Kalman filter parameter initialization requires a certain number of filtering times, the localization results and the calculation results of removing the first five groups of data by localization error are shown in Figures 16 and 17.

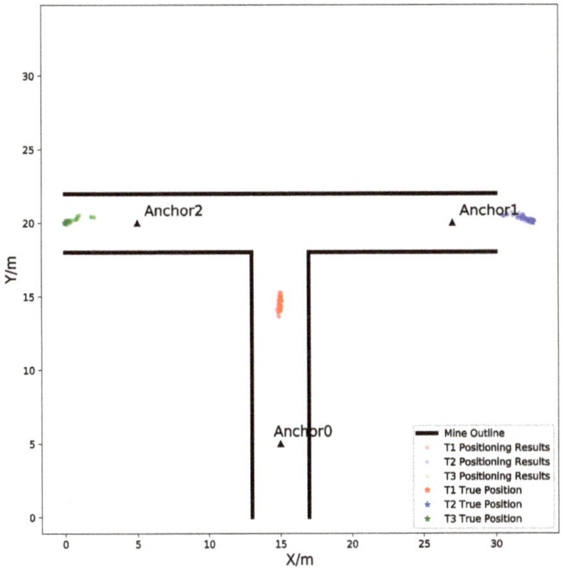

Figure 16. Localization results of the static experiment.

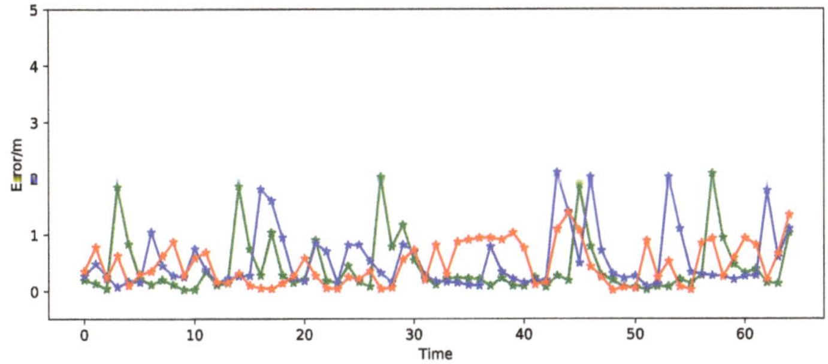

Figure 17. Error of localization results. T1 corresponds to the tag of the position of the red five-pointed star in the static experiment, T2 corresponds to the position of the blue five-pointed star, and T3 corresponds to the position of the green five-pointed star.

As can be seen from Figure 16, most of the localization results are beside the corresponding real positions. For Figure 17, we define the error as the distance between the real position and the position calculated by the localization system. For T1, T2, and T3, the average errors of our system localization are 0.47 m, 0.54 m, and 0.42 m, respectively. The reason why the error reaches 1 M or even 2 m is that NLOS is wrongly identified as LOS. The error in ranging cannot be eliminated in the linear model, so the error will be transferred to the positioning process in the subsequent process, resulting in a large location error. As T2 and T3 are located on one side of all anchors, the distance errors cannot cancel each other, and their location errors are larger than T1.

Figure 18 shows the results of the dynamic experiment. Since the real position of the tag at each moment could not be determined during the dynamic experiment, we could not calculate the error and could only compare the real track of tag with the track formed by the location result of our location system. As can be seen from Figure 18, the location at the first few moments has a large error, which is mainly due to the slow initialization of the Kalman filter parameters. When the parameters of the Kalman filter are gradually stabilized, the localization accuracy of the system begins to improve, and the movement track of the tag calculated by the system is almost consistent with that of the real tag. The reason for the large deviation of the location track lies in the error of the NLOS/LOS classification of the tag and one of the three anchors, which leads to the location deviation.

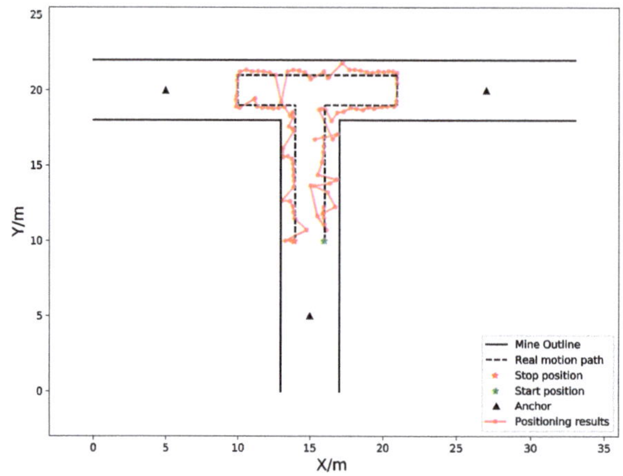

Figure 18. The dynamic tracking experiment.

According to the results of static and dynamic experiments, the accuracy of our localization system can reach decimeter level in complex tunnel environment. It can not only get the position of motionless personnel, but also track and locate the moving personnel in real time. When the line-of-sight condition is occasionally misidentified, the positioning result will have a large error. However this error is rare and permitted in the coal mine environment.

6. Conclusions

In this paper, we develop a personnel localization system for coal mines. On the one hand, we propose to use frame transmission diagnostic data for line-of-sight classification. From the classification results, CNN model has high classification accuracy for our data and can be used as our NLOS/LOS classifier. Compared with MLP, it has a higher f1-score, higher classification accuracy, and better performance for NLOS. On the other hand, we design a complete localization system, including the calculations used from data acquisition to the calculation of location results. From the results of static and dynamic experiments, it can be seen that our system has good localization performance for static or moving tags. For static tag, localization accuracy of half a meter can be achieved; For a moving tag, we can track the movement of the tag well. However, in different tunnel environments, different ranging environments will lead to different signals, so the robustness for different environments is low. In out future work, we will focus on the concurrency problem when the tag capacity is large.

Author Contributions: Conceptualization, Y.Z.; methodology, Y.Z.; software, Y.Z.; validation, M.W.; formal analysis, M.W.; investigation, M.W.; resources, M.W.; data curation, Y.Z.; writing—original draft preparation, M.W. and Y.Z.; writing—review and editing, M.W. and Y.Z.; project administration, M.W.; funding acquisition, M.Z. All authors have read and agreed to the published version of the manuscript.

Funding: This research was funded by the National Natural Science Foundation of China; grant number 61701237.

Acknowledgments: The authors would like to thank J.Y. and J.W. at the Department of Mechanical Engineering of NanJing University of Science and Technology for their contribution to the field experiment work.

Conflicts of Interest: The authors declare no conflict of interests.

References

1. Choliz, J.; Eguizabal, M.; Hernandezsolana, A.; Valdovinos, A. Comparison of Algorithms for UWB Indoor Location and Tracking Systems. In Proceedings of the 2011 IEEE 73rd Vehicular Technology Conference (VTC Spring), Budapest, Hungary, 15–18 May 2011 .
2. Cheng, L.; Zhao, A.; Wang, K.; Li, H.; Chang, R. Activity Recognition and Localization based on UWB Indoor Positioning System and Machine Learning. In Proceedings of the 2020 11th IEEE Annual Information Technology, Electronics and Mobile Communication Conference (IEMCON), Vancouver, BC, Canada, 4–7 November 2020.
3. Li, B.; Zhao, K.; Saydam, S.; Rizos, C.; Wang, J.; Wang, Q. *Third Generation Positioning System for Underground Mine Environments: An Update on Progress*; Far East Journal of Electronics and Communications (FJEC): Prayagraj, India, 2018.
4. Aca, B.; Pf, B.; Pmt, B. UWB-based sensor networks for localization in mining environments. *Ad Hoc Netw.* **2009**, *7*, 987–1000.
5. Capraro, F.; Segura, M.; Sisterna, C. Human Real Time Localization System in Underground Mines using UWB. *IEEE Lat. Am. Trans.* **2020**, *18*, 392–399. [CrossRef]
6. Zheng, X.; Wang, B.; Zhao, J. High-precision positioning of mine personnel based on wireless pulse technology. *PLoS ONE* **2019**, *14*, e0220471. [CrossRef] [PubMed]
7. Bao, J.; Wang, H. Research on Key Technologies of High-Precision Location Based Service System for Intelligent Mines. ResearchGate.PrePrint. Version 1. Posted 17 December 2020. Available online: https://www.sciencegate.app/app/document/download#10.21203/rs.3.rs-127354/v1 (accessed on 29 May 2022).
8. Ming, L. A Novel Ultra-Wideband Hybrid Localization Scheme in Coal Mine. *J. Commun.* **2015**, *10*, 889–895.
9. Yoon, J.; Kim, H.; Seo, D.; Nam, H. Performance Comparison of NLOS Detection Methods in UWB. In Proceedings of the 2021 International Conference on Information and Communication Technology Convergence (ICTC), Jeju Island, Korea, 19–21 October 2021; pp. 1486–1489. [CrossRef]
10. Liu, Q.; Yin, Z.; Zhao, Y.; Wu, Z.; Wu, M. UWB LOS/NLOS identification in multiple indoor environments using deep learning methods. *Phys. Commun.* **2022**, *52*, 101695. [CrossRef]
11. Altstidl, T.; Kram, S.; Herrmann, O.; Stahlke, M.; Feigl, T.; Mutschler, C. Accuracy-Aware Compression of Channel Impulse Responses using Deep Learning. In Proceedings of the 2021 International Conference on Indoor Positioning and Indoor Navigation (IPIN), Lloret de Mar, Spain, 29 November–2 December 2021; pp. 1–8. [CrossRef]
12. Song, B.; Li, S.L.; Tan, M.; Ren, Q.H. A fast imbalanced binary classification approach to NLOS identification in UWB positioning. *Math. Probl. Eng.* **2018**, *2018*, 1580147. [CrossRef]
13. Fan, R.; Du, X. NLOS Error Mitigation Using Weighted Least Squares and Kalman Filter in UWB Positioning. *arXiv* **2022**, arXiv:2205.05939.
14. Liu, R.; Yuen, C.; Do, T.N.; Jiao, D.; Liu, X.; Tan, U.X. Cooperative Relative Positioning of Mobile Users by Fusing IMU Inertial and UWB Ranging Information. In Proceedings of the 2017 IEEE International Conference on Robotics and Automation (ICRA), Singapore, 29 May–3 June 2017.
15. Savic, V.; Ferrer-Coll, J.; Angskog, P.; Chilo, J.; Stenumgaard, P.; Larsson, E.G. Measurement Analysis and Channel Modeling for TOA-Based Ranging in Tunnels. *IEEE Trans. Wirel. Commun.* **2015**, *14*, 456–467. [CrossRef]
16. Jiang, C.; Shen, J.; Chen, S.; Chen, Y.; Liu, D.; Bo, Y. UWB NLOS/LOS Classification Using Deep Learning Method. *IEEE Commun. Lett.* **2020**, *24*, 2226–2230. [CrossRef]
17. Sang, C.L.; Steinhagen, B.; Homburg, J.D.; Adams, M.; Rückert, U. Identification of NLOS and Multi-Path Conditions in UWB Localization Using Machine Learning Methods. *Appl. Sci.* **2020**, *10*, 3980. [CrossRef]
18. Tran, D.H.; Chung, B.; Jang, Y.M. GAN-based Data Augmentation for UWB NLOS Identification Using Machine Learning. In Proceedings of the 2022 International Conference on Artificial Intelligence in Information and Communication (ICAIIC), Jeju Island, Korea, 21–24 February 2022; pp. 417–420. [CrossRef]
19. Kaemarungsi, K.; Krishnamurthy, P. Modeling of indoor positioning systems based on location fingerprinting. In Proceedings of the IEEE INFOCOM, Hong Kong, China, 7–11 March 2004.
20. Trees, H.V.; Bell, K. On Geolocation Accuracy with Prior Information in Nonlineofsight Environment. In Proceedings of the EEE 56th Vehicular Technology Conference, Helsinki, Finland, 19–22 June 2002.

21. Poulose, A.; Han, D.S. Feature-Based Deep LSTM Network for Indoor Localization Using UWB Measurements. In Proceedings of the 2021 International Conference on Artificial Intelligence in Information and Communication (ICAIIC), Jeju Island, Korea, 20–23 April 2021; pp. 298–301. [CrossRef]
22. Djosic, S.; Stojanovic, I.; Jovanovic, M.; Djordjevic, G.L. Multi-algorithm UWB-based localization method for mixed LOS/NLOS environments. *Comput. Commun.* **2022**, *181*, 365–373. [CrossRef]
23. Creswell, A.; White, T.; Dumoulin, V.; Arulkumaran, K.; Sengupta, B.; Bharath, A.A. Generative Adversarial Networks: An Overview. *IEEE Signal Process. Mag.* **2018**, *35*, 53–65. [CrossRef]
24. Rosenblatt, F. *The Perceptron—A Perceiving and Recognizing Automaton*; Cornell Aeronautical Lab: Buffalo, NY, USA, 1957.
25. Shirinabadi, P.A.; Abbasi, A. UWB Channel Classification Using Convolutional Neural Networks. In Proceedings of the 2019 IEEE 10th Annual Ubiquitous Computing, Electronics & Mobile Communication Conference (UEMCON), New York, NY, USA, 10–12 October 2019.
26. Alavi, B. Distance Measurement Error Modeling for Time-of-Arrival-Based Indoor Geolocation. Ph.D. Thesis, Worcester Polytechnic Institute: Worcester, MA, USA, 2006.
27. Li, Q.; Li, R.; Ji, K.; Dai, W. Kalman Filter and Its Application. In Proceedings of the 2015 8th International Conference on Intelligent Networks and Intelligent Systems (ICINIS), Tianjin, China, 1–3 November 2015; pp. 74–77. [CrossRef]
28. Jimenez, A.R.; Seco, F. Comparing Decawave and Bespoon UWB location systems: Indoor/outdoor performance analysis. In Proceedings of the 2016 International Conference on Indoor Positioning and Indoor Navigation (IPIN), Alcala de Henares, Spain, 4–7 October 2016.

applied sciences

MDPI

Article

A Two-Objective ILP Model of OP-MATSP for the Multi-Robot Task Assignment in an Intelligent Warehouse

Jianqi Gao, Yanjie Li *, Yunhong Xu and Shaohua Lv

School of Mechanical Engineering and Automation, Harbin Institute of Technology (Shenzhen), Shenzhen 518071, China; gaojianqi205a@stu.hit.edu.cn (J.G.); 19s053099@stu.hit.edu.cn (Y.X.); 19S053101@stu.hit.edu.cn (S.L.)

* Correspondence: autolyj@hit.edu.cn

Abstract: Multi-robot task assignment is one of the main processes in an intelligent warehouse. This paper models multi-robot task assignment in an intelligent warehouse as an open-path multi-depot asymmetric traveling salesman problem (OP-MATSP). A two-objective integer linear programming (ILP) model for solving OP-MDTSP is proposed. The theoretical bound on the computational time complexity of this model is $O(n!)$. We can solve the small multi-robot task assignment problem by solving the two-objective ILP model using the Gurobi solver. The multi-chromosome coding-based genetic algorithm has a smaller search space, so we use it to solve large-scale problems. The experiment results reveal that the two-objective ILP model is very good at solving small-scale problems. For large-scale problems, both EGA and NSGA3 genetic algorithms can efficiently obtain suboptimal solutions. It demonstrates that this paper's multi-robot work assignment methods are helpful in an intelligent warehouse.

Keywords: multi-robot task assignment; intelligent warehouse; OP-MATSP; ILP; genetic algorithm

1. Introduction

With the development of the e-commerce and logistics industries, the advantages of the intelligent warehouse have begun to emerge [1]. The intelligent warehouse is mainly deployed by the multi-robot system [2], which can significantly improve efficiency and reduce cost. Amazon's order fulfillment center is the most mature intelligent warehouse globally [3]. As shown in Figure 1, the intelligent warehouse comprises three parts: robots, inventory pods, and inventory stations. The robots can operate autonomously in narrow passages. All robots are homogeneous, where every robot has the same ability to move the inventory pod. Robots are sometimes referred to as automated guided vehicles (AGVs). The inventory pods are containers that can hold one or more kinds of goods and can be transported by robots. The inventory stations have two types: picking stations and replenishment stations. Picking or replenishing goods from inventory pods is primarily done by workers. The workflow of an intelligent warehouse mainly includes multi-robot task assignment [4] and path-finding [5]. The two processes interact with each other [6]. Practical task assignments can reduce costs and avoid path conflicts.

There are many methods for solving multi-robot task assignment problems, including combinatorial optimization methods, market-based methods, swarm intelligence methods, clustering, and so on [7]. Multi-robot task assignment also has many applications in industrial manufacturing, especially flexible manufacturing systems [8]. In [9], the scholars explored the task assignment of material handling by multiple AGVs in complex random production lines. In [10], the scholars use the tabu search algorithm to solve the problem of simultaneous task assignments between machines and multiple AGVs in flexible manufacturing systems. In [11], the scholars discuss multi-objective multi-AGV task assignment in flexible manufacturing systems using evolutionary algorithms. However, few studies are about multi-robot task assignment in an intelligent warehouse.

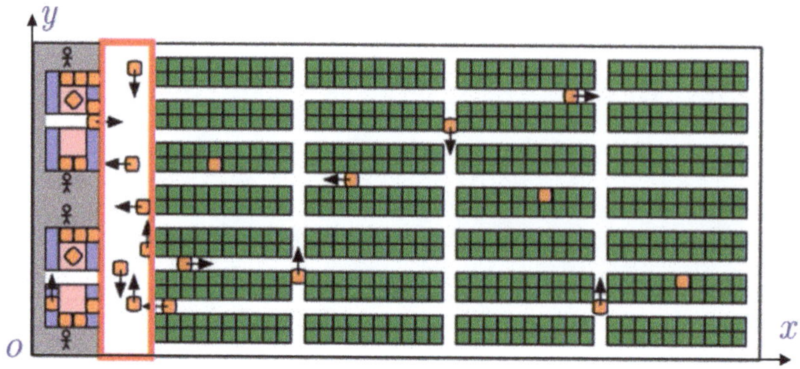

Figure 1. Two-dimensional schematic of Amazon's order fulfillment center [3]. The orange squares represent the robots. The green squares represent the inventory pods. The inventory stations are shown only on the left side and are usually located around the perimeter of warehouse.

According to the number of tasks assigned during every time window, task assignment can be divided into single-task assignment (STA) and multi-task assignment (MTA) [12]. Single-task assignment (STA) is deployed quickly, but multi-task assignment (MTA) is more helpful in reducing costs and increasing efficiency. Single-task assignment (STA) is also known as online task assignment. When a task arrives, the system assigns it to an idle robot based on some heuristics, such as the task order (first come and first served, FCFS) [13], the task priority [14], the distance between the task and the idle robot, the utilization of totes, the age of the tasks [15], and so on. The system assigns multiple associated tasks to each robot through an optimization model during each time window in multi-task assignment (MTA). The scholars study multi-task assignment (MTA) in intelligent warehousing using the genetic algorithms in [8,16,17]. However, the above studies only analyze one kind of task, such as the picking task. They do not include all kinds of warehouse tasks, which simplifies the actual task assignment in an intelligent warehouse.

This paper's task assignment method belongs to multi-task assignment (MTA) and can assign three different kinds of tasks. As shown in Figure 2, during a specific time window, each robot starts from an initial position to complete a series of tasks and does not return. There are three kinds of tasks: replenishment, moving, and picking. The replenishment task means the robots carry the inventory pod to the replenishment station and return. The moving task means the robots carry the inventory pods from their initial position to another location. The picking task means the robots carry the inventory pods to the picking stations and return. We assume that the robots carry the inventory pods to the initial position when the robots complete the replenishment and picking task. The replenishment and picking tasks are considered to be node tasks. On the other hand, because the initial and end positions of the inventory pod in the moving task are different, the moving task is considered to be an arc task. Because of the arc tasks, the cost between node tasks and arc tasks varies depending on the order in which they are completed. The cost between the tasks is asymmetric. After completing a series of tasks, each robot does not return to the starting position but stays at the last task point and waits for the following command. According to the above, the intelligent warehouse's multi-robot task assignment problem can be considered an open-path multi-depot asymmetric traveling salesman problem (OP-MATSP). We treat each task as a city node and each robot as a traveling salesman. Compared to earlier research, this work is the first to propose the OP-MATSP.

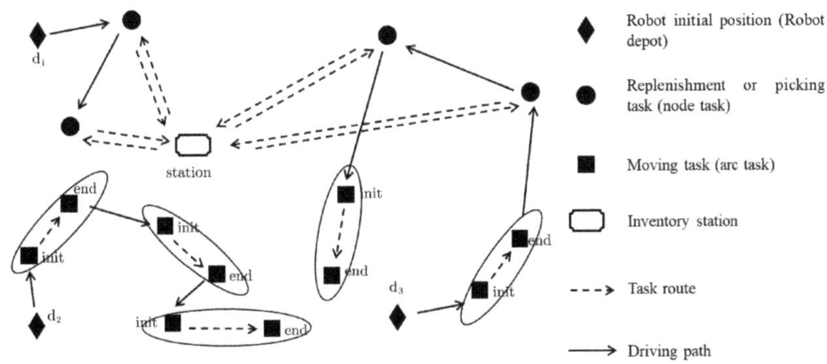

Figure 2. The multi-robot task assignment in an intelligent warehouse.

Some scholars have studied warehouse task assignment based on the traveling salesman problem (TSP) [18], but they mainly discuss the order picking problem [19–21]. When an order is received, the system sends a worker or a robot from the station to pick the items of the order from different inventory pods and finally return to the station. We treat each good in the order as a city node, each robot as a traveling salesman, and the station as the depot. In these studies, each robot moves to the inventory pods to pick up all the goods in the order and then returns to the start position. The above studies all belong to the "worker-to-goods" model. This paper's task assignment is based on the "live to human" model, which is more efficient in an intelligent warehouse.

For the optimization objective of the problem, the studies in [19–21] only design the pickup route for each vehicle through TSP, shortening its own distance to fulfill the order. These studies do not consider the optimal total distance of all robots to complete the orders. In [10], the optimization objective is mainly to minimize the manufacturing time to ensure the system's operating efficiency without balancing other optimization objectives. In order to reduce the amount of charging and replacing, some scholars take the device's energy efficiency as an optimization objective. The scholars in [22] propose a distributed resource management mechanism to determine the optimal transmission power for each device, in which the interests, physical relationships, and energy availability among devices are all considered. In this paper, we assume that the power of each robot is infinite. This paper optimizes the total distance and total time for all robots to complete the task simultaneously. The total distance is also known as the sum-of-cost (SOC). The total time is known as the makespan (MS).

OP-MATSP can be converted from the multi-depot traveling salesman problem (MTSP) [23]. As shown in Figure 3, each starting depot only has one traveling salesman who departs to visit the city node in the MTSP. When all city nodes have been visited, all traveling salesmen return to their depots. The cost between each city node is the same. The cost between each task node in this paper is asymmetric. By adding asymmetric restrictions, such as [24], we can modify the MTSP model to a multi-depot asymmetric traveling salesman problem (MATSP) model. In general, asymmetric models are more complex to solve than symmetric models [25]. In this paper, each robot stops at the last task node after completing a series of tasks and does not return to the starting depot, which is an open path problem. By eliminating the last edge of the traveling salesman returning to the starting depot, we can transfer the MATSP model to the OP-MATSP.

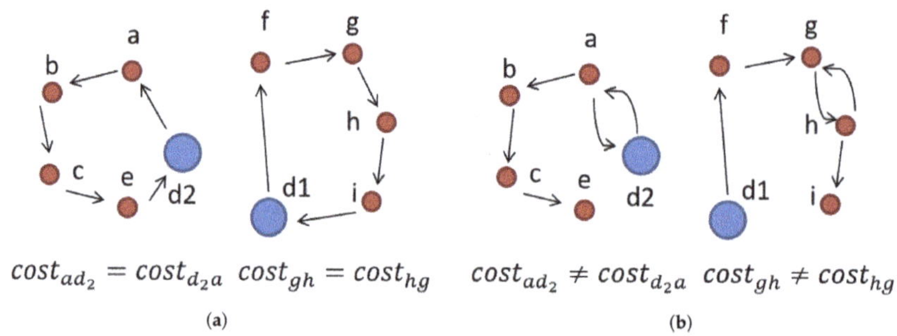

Figure 3. The diagram of MTSP and OP-MATSP. (**a**) Multi-depot traveling salesman problem (MTSP). (**b**) Open path multi-depot asymmetric traveling salesman problem (OP-MATSP). d_1 and d_2 represent depot. a to c, e to i represent the city nodes that the traveling salesman will visit.

As a generalization of TSP, the OP-MATSP is also an NP-hard combinatorial optimization problem. The exact same algorithms can be used for OP-MATSP. We firstly build a two-objective integer programming (ILP) model of OP-MATSP. Then, we solve the small-scale problems using the Gurobi solver [26]. The Gurobi solver is the fastest and most powerful mathematical solver available for ILP problems. To solve its non-inferior solution, we transform the two-objective integer linear programming (ILP) model into a single-objective problem by the linear weighting method [27]. We assign weight ω_1 and ω_2 to each objective function, and then add them together to form a new objective function and obtain a solution to the two-objective integer linear programming (ILP) model by solving this new objective function. However, it becomes complicated to solve the large-scale problem. In this paper, the multi-chromosome coding-based genetic algorithms [28,29] are used to solve the large-scale problems. The multi-chromosome coding genetic algorithm has been proven to have a smaller search space and faster search speed in solving the multiple traveling salesman problems (mTSP) [30]. We treat each robot's task sequence as a chromosome and each task as a gene in the chromosome. GEATPY is a high-performance genetic algorithm library. In this paper, the multi-chromosome coding-based genetic algorithms are selected in this library to solve the large-scale multi-robot task assignment problem. We first solve small-scale problems using the multi-chromosome coding-based genetic algorithm in GEATPY and compare the results with the Gurobi solver. We then use the algorithms with higher solution performance to solve large-scale problems.

The main contributions of this paper can be summarized as follows:

- This paper regards the multi-robot task assignment in an intelligent warehouse as an open-path multi-depot asymmetric traveling salesman problem (OP-MATSP). Moreover, the OP-MATSP is proposed for the first time in this paper.
- A two-objective integer linear programming (ILP) model of OP-MATSP is established. Through this ILP model, we solve small-scale multi-robot task assignment problems using the Gurobi solver.
- Multi-chromosome coding-based genetic algorithms are implemented to solve large-scale multi-robot task assignment problems.

The remainder of this paper is structured as follows. Section 2 gives the problem description and the two-objective ILP model. Section 3 presents the small-scale and large-scale experiments' results and discussions. Section 4 provides the conclusions.

2. Methods

2.1. Task Definition

A warehouse task is determined by the initial position of the inventory pod, the end position of the inventory pod, and the position of the picking (replenishment) station. We use a six-dimensional coordinate to define the warehouse task as:

$$t = (x_{init}, y_{init}, x_{end}, y_{end}, x_{station}, y_{station}) \tag{1}$$

where x_{init}, y_{init} represents the initial position of the inventory pod, x_{end}, y_{end} represents the end position of the inventory pod, and $x_{station}, y_{station}$ represents the position of the picking or replenishment station.

We define two kinds of task cost: task own cost (TOC) and task-associated cost (TAC). TOC is the cost for the robot to complete a task. TAC is the cost for a robot to go to another task from the current task. The distance a robot travels to complete a task can be seen as the task cost. In an intelligent warehouse, the robot can only move straight in the narrow passages between the inventory pods, so we can use the Manhattan distance to calculate the task cost. Then TOC of the node task and arc task can be defined as:

$$\begin{aligned} TOC(t_{node}) = 2|x_{init} - x_{station}| \\ + 2|y_{init} - y_{station}| \end{aligned} \tag{2}$$

$$TOC(t_{arc}) = |x_{init} - x_{end}| + |y_{init} - y_{end}| \tag{3}$$

The TAC from task t to task t' is the Manhattan distance between the end position of task t and the initial position of another task t'. It can be defined as:

$$TAC(t, t') = |x_{end} - x'_{init}| + |y_{end} - y'_{init}| \tag{4}$$

where t and t' can represent the node task and arc task.

2.2. Multi-Robot Task Assignment in an Intelligent Warehouse

Figure 4 is a diagram of task sequences for the multi-robot task assignment in an intelligent warehouse. There are 5 robots and 15 tasks, including 10 node tasks and 5 arc tasks. The task sequences of robots are $\langle r_1, 12, 4, 13, 9\rangle$, $\langle r_2, 1, 14\rangle$, $\langle r_3, 6, 8, 2\rangle$, $\langle r_4, 7, 11\rangle$, and $\langle r_5, 10, 5, 0, 3\rangle$. Each robot starts from a different position to complete tasks and does not return to its start position. Each task can be completed only once by one robot. Each task has a unique TOC. The TAC between node task t and arc task t' is closely related to the task execution order, that is, $TAC(t, t') \neq TAC(t', t)$. This paper contains two optimization objectives: sum-of-cost and makespan. The sum-of-cost is the sum of TOC and TAC for all tasks. Makespan is the maximum value of the sum of TAC and TOC for the tasks in each sequence. In a specific time window, when the number and type of all tasks are confirmed, the $\sum TOC$ is a constant. Therefore, the value of sum-of-cost is mainly determined by the TAC of all tasks, while makespan is related to the TOC and TAC of the tasks in each sequence.

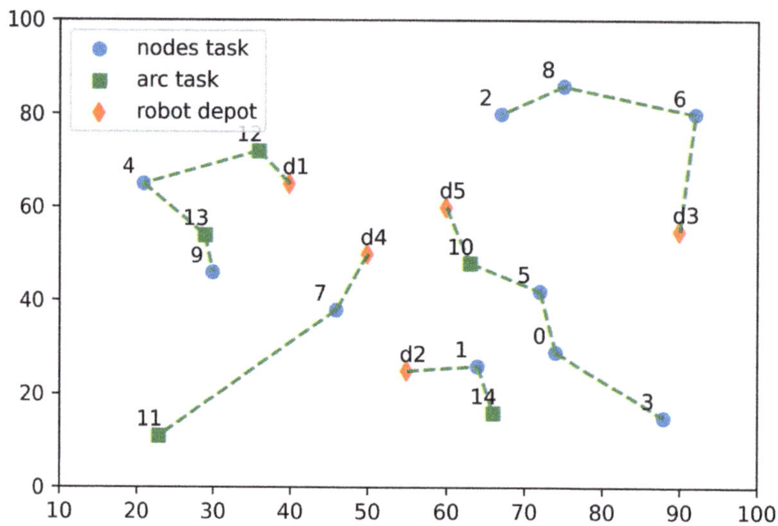

Figure 4. Task sequences for the multi-robot task assignment.

2.3. Open-Path Multi-Depot Asymmetric Traveling Salesman Problem (OP-MATSP)

We can model the multi-robot task assignment in an intelligent warehouse as an open-path multi-depot asymmetric traveling salesman problem (OP-MATSP). Each task is considered to be a city node, each robot is considered to be a traveling salesman, and the initial position of a robot is considered to be a starting depot.

OP-MATSP is another variant of TSP, consisting of m traveling salesmen, m depots, and n city nodes. Only one traveling salesman can visit each city node. The mth traveling salesman starts from the mth depot to visit respective city nodes until m traveling salesmen visit all city nodes. Each traveling salesman stays at the last city node without returning to the starting depot. OP-MATSP can be described as a directed weighted graph $G = (V, A)$ with set $V = V_n \cup V_m$ and set A. $V = \{1, \cdots, n, n+1, \cdots, n+m\}$ is the city node and depot set. $A = \{(i,j)|i,j \in V, i \neq j\}$ is the arc set. Subset $V_n = \{1, \cdots, n\}$ represents n city nodes. Subset $V_m = \{n+1, \cdots, n+m\}$ represents m depots. $C = (c_{ij})$ is the cost matrix associated with each arc $(i,j) \in A$, and $c_{ij} \neq c_{ji}$ (asymmetric).

2.4. Formulation of the Two-Objective ILP Model

In this section, a two-objective integer linear programming (ILP) model of OP-MATSP is established. We first propose the sets, parameters, decision variables, constraints, and objective function of the two-objective integer linear programming (ILP) model for the multi-depot asymmetric traveling salesman problem (MATSP). Then, we can give the two-objective ILP model for OP-MATSP by changing the objective function.

(1) Sets
V_n: the city node set.
V_m: the depot node set. Since each traveling salesman starts from a unique depot node, we assume that the set of traveling salesmen is equal to the depot node set V_m.
V: the city and depot node set.

(2) Parameters
c_{ij}: the cost from node i to node j.
q_i: the own cost of city node i.

(3) Decision variables
x_{ij}^m: if mth traveling salesman visits the node i and moves to the next node j, then $x_{ij}^m = 1$; otherwise, $x_{ij}^m = 0, i, j \in V, m \in V_m$.

y_i^m: if mth traveling salesman visits the node i, then $y_i^m = 1$; otherwise, $y_i^m = 0$, $i \in V$, $m \in V_m$.

(4) Constraints

(a) The mth traveling salesman can only visit the city node from the mth depot node, but not other depot nodes. To guarantee the above assumptions, we set the following constraints:

$$x_{ij}^m = x_{ji}^m = 0, \forall i, j \in V_m, i \neq j, \forall m \in V_m \tag{5}$$

(b) To prevent the traveling salesman from repeatedly visiting the same city node or depot node, the following constraint is established:

$$x_{ii}^m = 0, \forall i \in V, \forall m \in V_m \tag{6}$$

(c) The relationship between decision variables x_{ij}^m and y_i^m can be expressed as:

$$\sum_{j \in V} x_{ji}^m = y_i^m, \forall i \in V_n, \forall m \in V_m \tag{7}$$

$$\sum_{j \in V} x_{ij}^m = y_i^m, \forall i \in V_n, \forall m \in V_m \tag{8}$$

(d) The number of node sequences is the same as the number of traveling salesmen. To ensure each traveling salesman visits only one node sequence, we add the following constraints:

$$\sum_{j \in V_n} x_{ij}^m = 1, \forall i, m \in V_m \tag{9}$$

$$\sum_{j \in V_n} x_{ji}^m = 1, \forall i, m \in V_m \tag{10}$$

(e) The city node can only be visited by one traveling salesman. To prevent one city node from being repeatedly visited by different traveling salesmen, the following constraints are established:

$$\sum_{j \in V} \sum_{m \in V_m} x_{ij}^m = 1, \forall i \in V_n, i \neq j \tag{11}$$

$$\sum_{i \in V} \sum_{m \in V_m} x_{ij}^m = 1, \forall j \in V_n, i \neq j \tag{12}$$

(f) To prevent the traveling salesman from visiting the previous node after visiting the current node, we establish the following constraints:

$$\sum_{m \in V_m} x_{ij}^m + \sum_{m \in V_m} x_{ji}^m < 2, \forall i, j \in V, i \neq j \tag{13}$$

(g) To ensure the continuity of node sequence and avoid the fracture of node sequence, the following flow balance constraint is established:

$$\sum_{j \in V} x_{ij}^m = \sum_{j \in V} x_{ji}^m, \forall i \in V, \forall m \in V_m \tag{14}$$

(h) We need to ensure that every node is visited and each traveling salesman's route forms a Hamiltonian cycle. To prevent sub-tours in the traveling salesman's routing, we have to add sub-tour elimination constraints (SECs). The SECs for the MATSP in this paper are Danzig–Fulkerson–Johnson (DFJ) [24]:

$$\sum_{i,j \in S} x_{ij} \leq |S| - 1, 2 < |S| \leq |V| - 1, \forall S \subset V \tag{15}$$

where $|S|$ represents the cardinality of subset $\mathcal{S}$, and $|V|$ represents the cardinality of set V. In each subset $\mathcal{S}$, sub-tours are prevented.

(i) To ensure that each traveling salesman starts from the depot node and visits at least one city node, we add the following constraints:

$$\sum_{i \in V_m} \sum_{j \in V_n} x_{ij}^m = 1, \forall m \in V_m \tag{16}$$

(5) Objective optimization functions

We will first propose two objective optimization functions based on MATSP. Sum-of-cost consists of two parts: the sum of the costs of all city nodes $\sum_{i \in V_n} q_i$ and the sum of the costs between nodes in the node sequence $\sum_{i \in V} \sum_{j \in V} \sum_{m \in V_m} c_{ij} x_{ij}^m$. Because $\sum_{i \in V_n} q_i$ is a constant, the expression of sum-of-cost (SOC) can be simplified as:

$$\min \quad f_1 = \sum_{i \in V} \sum_{j \in V} \sum_{m \in V_m} c_{ij} x_{ij}^m \tag{17}$$

Makespan is the maximum value of the total cost of each node sequence. Makespan also includes two parts: the sum of the costs of all city nodes in the sequence $\sum_{i \in V} \sum_{j \in V} c_{ij} x_{ij}^m$ and the sum of the costs between nodes in the node sequence $sum_{i \in V_n} q_i y_i^m$. We define the makespan as:

$$\min \quad f_2 = \max_{m \in V_m} \left\{ \sum_{i \in V} \sum_{j \in V} c_{ij} x_{ij}^m + \sum_{i \in V_n} q_i y_i^m \right\} \tag{18}$$

Finally, we obtain the objective optimization function of OP-MATSP by changing the two objective functions of MATSP. The difference between OP-MATSP and MATSP is whether all traveling salesmen return to the depot after visiting all city nodes. Therefore, we can get two objective optimization functions of OP-MATSP by subtracting the cost of the traveling salesman returning the depot from the last city node in Equations (17) and (18).

The sum-of-cost of OP-MATSP can be expressed as:

$$\min \quad f_1 = \sum_{i \in V} \sum_{j \in V} \sum_{m \in V_m} c_{ij} x_{ij}^m - \sum_{i \in V_n} \sum_{j \in V_m} \sum_{m \in V_m} c_{ij} x_{ij}^m \tag{19}$$

where $\sum_{i \in V_n} \sum_{j \in V_m} \sum_{m \in V_m} c_{ij} x_{ij}^m$ represents the cost from the last city node to the depot of all traveling salesmen.

The makespan of OP-MATSP can be expressed as:

$$\min \quad f_2 = \max_{m \in V_m} \left\{ \sum_{i \in V} \sum_{j \in V} c_{ij} x_{ij}^m - \sum_{i \in V_n} \sum_{j \in V_m} c_{ij} x_{ij}^m + \sum_{i \in V_n} q_i y_i^m \right\}$$

where $\sum_{i \in V_n} \sum_{j \in V_m} c_{ij} x_{ij}^m$ represents the cost of the mth traveling salesman routing from the last city node to the depot.

(6) The two-objective ILP model

In conclusion, a two-objective ILP model of OP-MATSP is established, through which we can solve the multi-robot task assignment in an intelligent warehouse. The entire two-objective ILP model is:

$$\min \quad f_1 = \sum_{i \in V} \sum_{j \in V} \sum_{m \in V_m} c_{ij} x_{ij}^m - \sum_{i \in V_n} \sum_{j \in V_m} \sum_{m \in V_m} c_{ij} x_{ij}^m \tag{20}$$

$$\min \quad f_2 = \max_{m \in V_m} \left\{ \sum_{i \in V} \sum_{j \in V} c_{ij} x_{ij}^{mk} - \sum_{i \in V_n} \sum_{j \in V_m} c_{ij} x_{ij}^m + \sum_{i \in V_n} q_i y_i^m \right\} \tag{21}$$

$$\text{s.t.} \quad \sum_{j\in V_n} x_{ij}^m = 1, \forall i, m \in V_m \tag{22}$$

$$\sum_{j\in V_n} x_{ji}^m = 1, \forall i, m \in V_m \tag{23}$$

$$\sum_{j\in V} \sum_{m\in V_m} x_{ij}^m = 1, \forall i \in V_n, i \neq j \tag{24}$$

$$\sum_{i\in V} \sum_{m\in V_m} x_{ij}^{mk} = 1, \forall j \in V_n, i \neq j \tag{25}$$

$$x_{ij}^m = x_{ji}^m = 0, \forall i, j, m \in V_m, i \neq j \tag{26}$$

$$x_{ii}^m = 0, \forall i \in V, \forall m \in V_m \tag{27}$$

$$\sum_{m\in V_m} x_{ij}^m + \sum_{m\in V_m} x_{ji}^m \leq 2, \forall i, j \in V, i \neq j \tag{28}$$

$$\sum_{j\in V} x_{ij}^m = \sum_{j\in V} x_{ji}^m, \forall i \in V, \forall m \in V_m \tag{29}$$

$$\sum_{i,j\in S} x_{ij} \leq |S| - 1, 2 < |S| \leq |V| - 1, \forall S \subset V \tag{30}$$

$$\sum_{i\in V_m} \sum_{j\in V_n} x_{ij}^m = 1, \forall m \in V_m \tag{31}$$

$$\sum_{j\in V} x_{ji}^m = y_i^m, \forall i \in V_n, \forall m \in V_m \tag{32}$$

$$\sum_{j\in V} x_{ij}^m = y_i^m, \forall i \in V_n, \forall m \in V_m \tag{33}$$

2.5. Analysis of Computational Time Complexity

Before computing the above ILP model numerically, we first perform a theoretical analysis of the OP-MATSP's computational time complexity bound. We use the example of OP-MATSP in Figure 3b to analyze the computational time complexity bound. There are two depots, two traveling salesmen, and eight city nodes. Each depot has a traveling salesman to visit the city node. As shown in Figure 5, the traveling salesman of depot d_1 chooses the first city node to visit from the eight city nodes, then the traveling salesman of depot d_2 chooses another city node to visit from the remaining seven city nodes, until the two traveling salesmen have visited all city nodes. We can see that the bound of the computational time complexity in this example is $O(8!)$. We can then can infer that the bound of the computational time complexity of an OP-MATSP with m traveling salesmen and n city nodes is $O(n!)$.

From the computational time complexity of OP-MATSP, we can see that the computational cost of our proposed integer linear programming model is acceptable for small-scale multi-robot task assignment problems but unacceptable for large-scale problems.

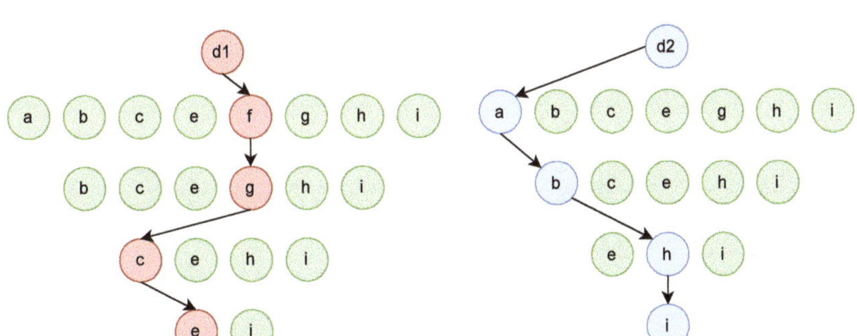

Figure 5. The diagram of all possible state space to search. d_1 and d_2 represent depot. a to c, e to i represent the city nodes that the traveling salesman will visit.

3. Results and Discussion

We use Python3 for coding. The computer's configuration is as follows: Intel Core i7 3.0 GHz processor, 32 GB RAM, and Windows 10 operating system.

3.1. Setting and Instance Generation

The coordinates of the tasks and depots in the small-scale problems are randomly generated within the scope of 25 × 16 (m × m), while the coordinates in the large-scale problems are randomly generated within the scope 1000 × 1000 (m × m). For the small-scale problem, there are four experiment instances. They include 3 robots with 10 tasks, 3 robots with 15 tasks, 5 robots with 20 tasks, and 5 robots with 25 tasks. One experiment instance is conducted for a large-scale problem that consists of 5 robots with 100 tasks. The sum-of-cost (SOC) obtained in the following experiments does not contain the *TOC* of all tasks.

3.2. Small-Scale Problems Solved by Gurobi

This section uses the Gurobi solver to solve the small-scale multi-robot task assignment problem. First, we use the linear weighting method to give different weight coefficients w_1 and w_2 to the objective functions (19) and (20), respectively, to form the new objective function f_3:

$$\min \quad f_3 = w_1 \times f_1 + w_2 \times f_2 \tag{34}$$

when the weight coefficient (w_1, w_2) is equal to $(1, 0)$ and $(0, 1)$, we get the objective function with the minimum sum-of-cost (SOC) and makespan (MS). When $w_1, w_2 \in (0, 1)$, sum-of-cost (SOC) and makespan (MS) are both considered in the new objective function. Then, we use the Gurobi solver to solve the new ILP model. We set the weight coefficients (w_1, w_2) to $(1, 0)$ and $(0, 1)$ in four small-scale experiments, respectively. We set the maximum CPU runtime of the Gorubi solver to 900 s.

In order to verify the theoretical analysis of the computational time complexity in Section 2.5, we perform a statistical analysis of the CPU running time of the Gurobi solver in four small-scale experiments. Table 1 shows that as the scope of the experiment grows, the minimum SOC and minimum MS-based CPU runtimes grow longer. Furthermore, the CPU runtime based on minimum makespan (MS) is longer than that based on minimal sum-of-cost (SOC) in the same experiment. Figure 6 compares the CPU runtime based on the minimum SOC and the computational time complexity of the same experiment. The left and right vertical axes are the logarithmic values of CPU runtime and computational time complexity based on the minimum SOC, respectively. We can see that the growth trend of the above two curves is consistent with the increase of the problem size, which further verifies that our theoretical analysis of the computational time complexity is correct.

Figures 7–10 represent the task sequence of each robot in the small-scale experiment instances, respectively. Because the solution result based on the minimum sum-of-cost (SOC) is only related to the *TAC* between tasks and has nothing to do with the *TOC* of the task itself, the task nodes appear to cluster in Figures 7a–10a. The solution results obtained based on the minimum makespan (MS) are not only related to *TAC* but also *TOC*, so the task sequences are chaotic without the evident cluster phenomenon in Figures 7b–10b.

Table 1. The computer's CPU runtime and the computational time complexity.

Item	Ex-1	Ex-2	Ex-3	Ex-4
$CPUtime_{SOC}$	0.18	3.9	39.97	208.45
$CPUtime_{MS}$	1.08	362.08	900.57 [1]	900.4 [1]
$O(n!)$	10!	15!	20!	25!

[1] The CPU runtime reaches the set threshold of 900 s.

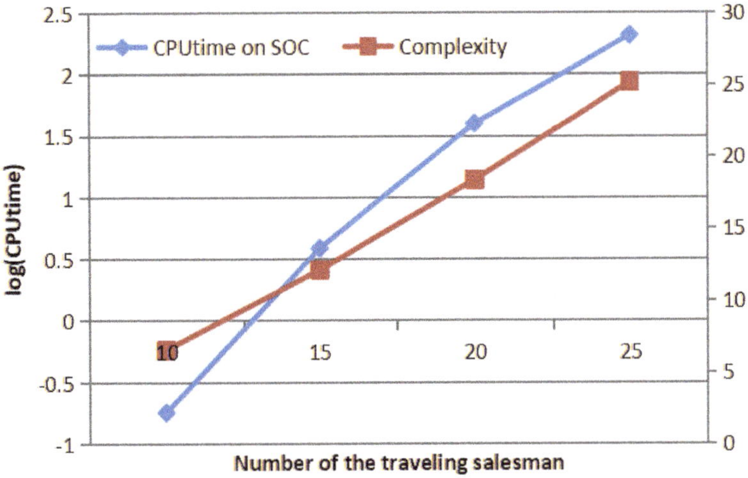

Figure 6. The comparison of the CPU runtime based on the minimum SOC and the computational time complexity.

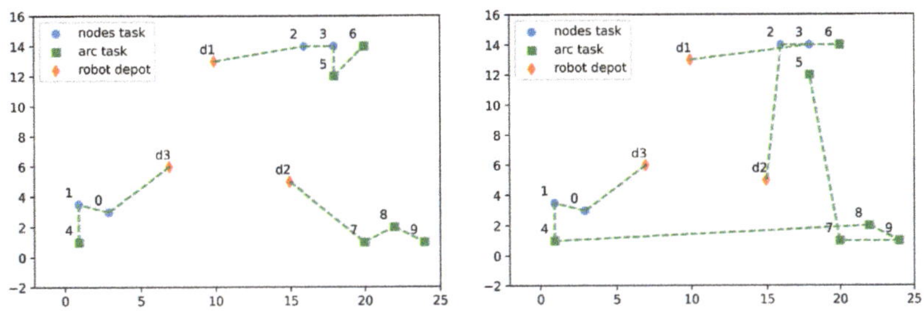

(**a**) Minimum sum-of-cost. (**b**) Minimum makespan.

Figure 7. Task sequences in experiment 1, including 3 robots and 10 tasks.

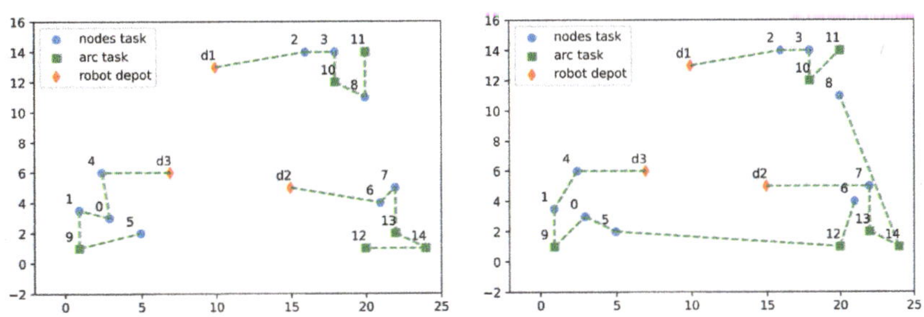

(**a**) Minimum sum-of-cost. (**b**) Minimum makespan.

Figure 8. Task sequences in experiment 2, including 3 robots and 15 tasks.

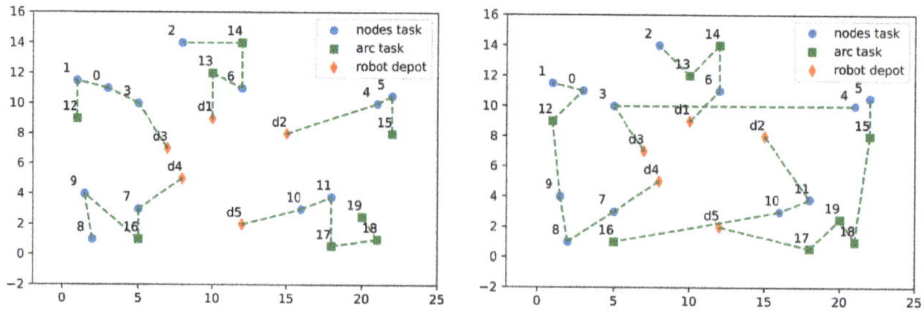

(**a**) Minimum sum-of-cost. (**b**) Minimum makespan.

Figure 9. Tasksequences in experiment 3, including 5 robots and 20 tasks.

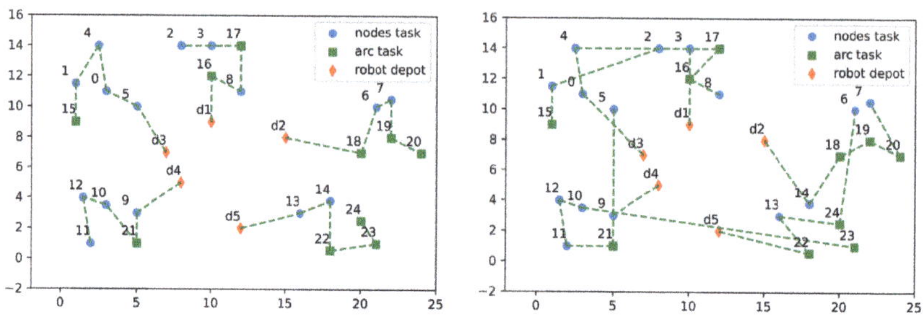

(**a**) Minimum sum-of-cost. (**b**) Minimum makespan.

Figure 10. Task sequences in experiment 4, including 5 robots and 25 tasks.

To explore the practical significance of the task assignment method in this paper, we compare it with single-task assignment (STA). Most of the existing intelligent warehouses adopt single-task assignment (STA). The approach based on the nearest distance between the start point of the task and the idle robot is more widely adopted. We now simulate it using the following method. First, we assign a task to the robot that is closest to the task starting position. Then, when a robot completes the task, we give the next task to the robot based on the nearest distance. When all tasks within the time window are given to all robots, we end the simulation process.

Based on the above four small-scale experiments, we compare the task assignment method based on the integer programming (ILP) model with the single-task assignment based on the nearest distance. In the nearest-distance-based single-task assignment, after the robot completes the task, it selects the nest task with the nearest distance to itself each time. Essentially, the single-task assignment gets a local optimal solution. In each task time window, the integer-linear-programming-based multi-task assignment (MTA) determines the execution order of tasks according to the relationship between tasks, which will obtain an optimal solution. We use Gap to represent the difference of sum-of-cost (SOC) or makespan (MS) obtained by single-task assignment (STA) based on the nearest distance and the ILP model. The expression of the Gap for sum-of-cost (SOC) and makespan (MS) is:

$$Gap_{SOC} = (SOC_{\mathrm{STA}} - SOC_{\mathrm{ILP}})/SOC_{\mathrm{STA}} \tag{35}$$

$$Gap_{MS} = (MS_{\mathrm{STA}} - MS_{\mathrm{ILP}})/MS_{\mathrm{STA}} \tag{36}$$

The results are shown in Table 2. The minimum sum-of-cost (SOC) and makespan (MS) obtained by the ILP model is reduced by at least 30% and 37.6%, respectively, compared to the nearest-distance-based single-task assignment (STA), and even by 44.3% and 50.2% in some cases. The results show that the integer linear programming (ILP)-based multi-task assignment (MTA) has great advantages over the single-task assignment method based on the nearest distance for small-scale problems.

Table 2. Comparison of single-task assignment (STA) and ILP-model-based task assignment for small-scale experiments.

Index	Item	Ex-1	Ex-2	Ex-3	Ex-4
SOC	STA	345	598.5	773	889.05
	ILP model	192	399	517.1	622.1
	Gap_{SOC}	44.3%	33.3%	33.1%	30%
MS	STA	135	229	227	277.6
	ILP model	77	143	113	139
	Gap_{MS}	43%	37.6%	50.2%	49.9%

To further explore the task assignment of each robot in the four small-scale experiments, the cost of each robot's task sequence is represented by a histogram under the patterns of $(w_1, w_2) = (1,0)$, $(w_1, w_2) = (0,1)$, and single-task assignment (STA). As shown in Figure 11, when $(w_1, w_2) = (0,1)$, the total cost of each robot's task sequence is relatively balanced and the value of makespan (MS) is the smallest. When $(w_1, w_2) = (1,0)$, the sum-of-cost (SOC) is the smallest. For the single-task assignment (STA) based on the nearest distance, the values of sum-of-cost (SOC) and makespan (MS) are larger than the other two task assignment patterns.

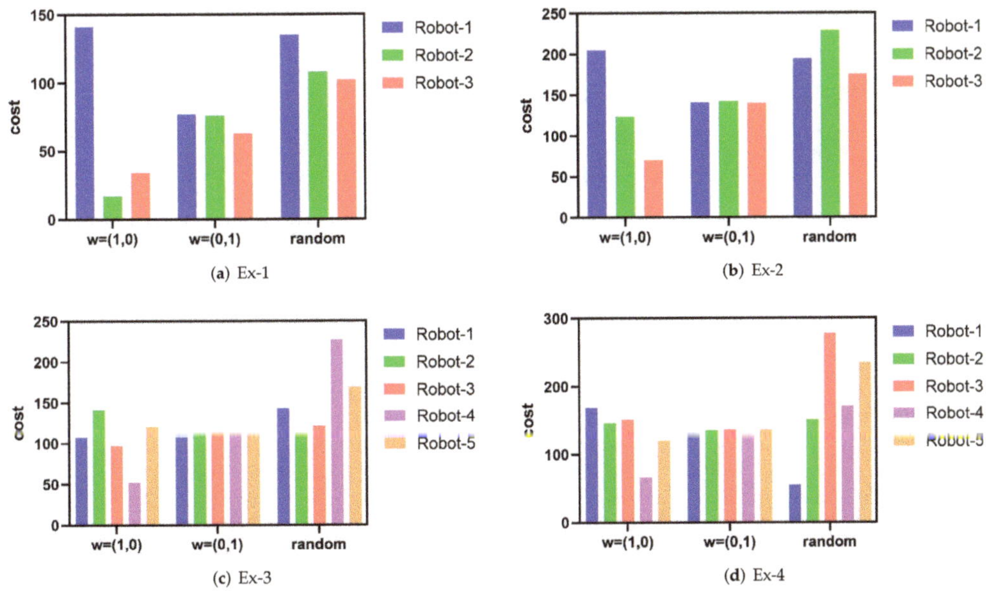

(a) Ex-1 (b) Ex-2 (c) Ex-3 (d) Ex-4

Figure 11. The cost for each robot's task sequence under different task assignment methods.

From the above analysis, we know that the two-objective integer linear programming (ILP)-based multi-robot task assignment in this paper has practical significance and can significantly improve the operating efficiency and reduce the operating cost of an intelligent warehouse.

3.3. Large-Scale Problems Solved by the Multi-Chromosome Coding-Based Genetic Algorithm

For large-scale problems, multi-chromosome coding-based genetic algorithms are implemented. This kind of genetic algorithm not only has a smaller search space [30] but also facilitates the integer coding for the multi-robot task assignment.

The multi-chromosome coding-based genetic algorithm also includes chromosome coding, population initialization, selection, crossover, mutation, and other processes. First, the integer coding process of the multi-chromosome genetic algorithm is described. The number of chromosomes of each individual in the population is equal to the number of robots. For a multi-robot task assignment problem containing n tasks and m robots, the task set is V_n, and the robot set is V_m. Each task represents a gene. Each individual in the population contains m chromosomes. The sequence of genes represents the order of tasks. After many generations of evolution, we pick out the best individuals from the population and assign them to robots. Then, 5 robots and 20 tasks are used to illustrate the multi-chromosome integer coding process, as shown in Figure 12. The specific generation process of a random initial population of size S is as follows: First, the order of the original task sequence needs to be disrupted, and then a robot is randomly chosen and assigned the first task of the new task sequence. It then randomly chooses a robot from all the robots and gives the second task to that robot. When all tasks are assigned, the individual of the population will be retained if every robot has at least one task. Otherwise, it will be discarded and re-assigned. The process is repeated until the population size reaches S. The objective function (34) is used to evaluate the population's individual. When weight coefficient (w_1, w_2) is equal to $(1, 0)$ and $(0, 1)$, respectively, we get the best individual of the population with the minimum sum-of-cost (SOC) and makespan (MS).

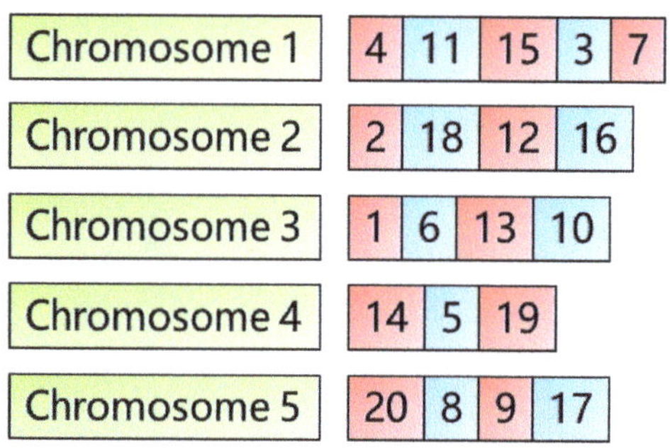

Figure 12. Multi-chromosome integer coding process of the genetic algorithm.

We use the multi-chromosome coding-based genetic algorithm in the GEATPY library to solve the large-scale multi-robot task assignment problem. GEATPY is a high-performance genetic algorithm library, which can be found on https://github.com/geatpy-dev/geatpy (accessed on 1 January 2022). We first test the eight kinds of multi-chromosome coding-based genetic algorithms on the small-scale experiment instances. Eight kinds of multi-chromosome coding-based genetic algorithms are shown in Table 3, including four types of single-objective and four types of multi-objective. Selection strategies of genetic algorithms include tournament, roulette wheel, and unconstrained random. We choose the partial matching crossover and reverse mutation operators that come with the GEATPY library for both crossover and mutation operators. We choose partial matching crossover and reverse mutation operators for both crossover and mutation operators. Regarding the crossover probability and mutation probability, we choose $P_c = 0.7$ and $P_m = 0.5$ in the single-objective genetic algorithm, respectively, and $P_c = 1$ and $P_m = 1$ in the multi-objective genetic algorithm. The iteration number of the genetic algorithm is 5000, and the population size is 100.

Then, we compare this with the results of the ILP model using the Gurobi solver, respectively. The experiment instances with different weight coefficients and different genetic algorithms are repeated 10 times, respectively. In each small-scale experiment, we statistically obtain the best and average values of sum-of-cost (SOC) and makespan (MS) from 10 repetitions of the genetic algorithms. We use Gap to represent the difference of sum-of-cost (SOC) or makespan (MS) obtained by the multi-chromosome coding-based genetic algorithm (GA) and the ILP model. The expression of Gap for sum-of-cost (SOC) and makespan (MS) is:

$$Gap_{SOC} = (SOC_{\mathrm{GA}} - SOC_{\mathrm{ILP}})/SOC_{\mathrm{GA}} \quad (37)$$

$$Gap_{MS} = (MS_{\mathrm{GA}} - MS_{\mathrm{ILP}})/MS_{\mathrm{GA}} \quad (38)$$

Table 3. Multi-chromosome coding-based genetic algorithm templets of GEATPY .

Type	Templet	Selection	Crossover	Mutation
Single-obj	soea_psy_EGA_templet [1]	Tournament	partial matching ($P_c = 0.7$)	Invertion ($P_m = 0.5$)
	soea_psy_SEGA_templet [2]	Tournament	partial matching ($P_c = 0.7$)	Invertion ($P_m = 0.5$)
	soea_psy_SGA_templet [3]	Roulette Wheel	partial matching ($P_c = 0.7$)	Invertion ($P_m = 0.5$)
	soea_psy_studGA_templet [31]	Tournament	partial matching ($P_c = 0.7$)	Invertion ($P_m = 0.5$)
Multi-obj	moea_psy_awGA_templet [4]	Tournament	partial matching ($P_c = 1$)	Invertion ($P_m = 1$)
	moea_psy_NSGA2_templet [32]	Tournament	partial matching ($P_c = 1$)	Invertion ($P_m = 1$)
	moea_psy_NSGA2_archive_templet [5]	Tournament	partial matching ($P_c = 1$)	Invertion ($P_m = 1$)
	moea_psy_NSGA3_templet [33]	unconstrained random	partial matching ($P_c = 1$)	Invertion ($P_m = 1$)

[1] This is the elitist reservation GA algorithm. [2] This is the strengthen elitist reservation GA algorithm. [3] This is the simple GA algorithm. [4] This is the multi-objective awGA algorithm. [5] This is the NSGA-II algorithm with global archive.

The comparison results of the multi-chromosome coding-based genetic algorithm and the ILP model are shown in Table 4. In the same experiment, the Pareto frontier extreme values obtained from the multi-objective genetic algorithm are equal to or close to the results obtained by the Gurobi solver. The best value of makespan obtained by awGA, NSGA2, NSGA2-archive, and NSGA3 in experiment 4 is smaller than the result obtained by the Gurobi solver. This indicates that when the problem scale reaches 5 robots and 25 tasks, the best value obtained by some genetic algorithms is better than that of the Gurobi solver.

To make it easier to compare the performance of each genetic algorithm in the four small-scale experiments, we use box-whisker plots to display the results. Based on Table 4, the box-whisker plots on the minimum sum-of-cost (SOC) and makespan (MS) are shown in Figures 13 and 14. It can be seen that as the number of robots and tasks increases, the Gap between the maximum and minimum results obtained by the same genetic algorithm becomes larger under the same number of iterations. The stability of the solution results gradually deteriorates. Among the eight tested genetic algorithms, the algorithms of EGA and NSGA3 perform relatively better. Therefore, we next choose these two algorithms to solve the large-scale problems.

The large-scale experiment instance includes 5 robots and 100 tasks. We use the two algorithms to solve the experimental case 10 times. To further explore the influence of the number of iterations of the genetic algorithm on the solution results, we set the number of iterations to 5000 and 20,000, respectively. The other parameters are the same as those of the small-scale experiment. We use the Gap to represent the difference between sum-of-cost (SOC) or makespan (MS) when the number of iterations is 5000 and 20,000. The expression of the Gap for sum-of-cost (SOC) and makespan (MS) is:

$$Gap_{SOC} = (SOC_{5000} - SOC_{20,000})/SOC_{5000} \quad (39)$$

$$Gap_{MS} = (MS_{5000} - MS_{20,000})/MS_{5000} \quad (40)$$

As shown in Table 5, when the number of iterations is 5000 and 20,000, the Gaps between the best value and the average values of the sum-of-cost (SOC) and makespan (MS) obtained by the EGA algorithm are all less than 4%. For the NSGA3 algorithm, the Gap is less than 1%. The results show that the two genetic algorithms have achieved good sub-optimal results when the number of iterations is 5000. They can be applied to other large-scale multi-robot task assignment problems of the intelligent warehouse. When the number of robots and tasks grows, we should appropriately increase the number of iterations of the genetic algorithm to obtain better results.

Table 4. Comparison results of the multi-chromosome coding-based genetic algorithms and ILP model in the small-scale experiments.

Ex	Index	soea-EGA		soea-SEGA		soea-SGA		soea-studGA		moea-awGA		moea-NSGA2-archive		moea-NSGA2		moea-NSGA3	
		Best	**Aveg**	**Best**	**Aveg**	**Best**	**Aveg**	**Best**	**Aveg**	**Best**	**Aveg**	**Best**	**Aveg**	**Best**	**Aveg**	**Best**	**Aveg**
Ex-1	SOC	**192**	195.8	**192**	200.1	195.5	204.6	**192**	196.7	**192**	198.2	**192**	194.6	**192**	196.5	**192**	**192**
	Gap_{SOC}	**0%**	1.94%	**0%**	4.05%	1.79%	6.16%	**0%**	2.39%	**0%**	3.13%	**0%**	1.34%	**0%**	1.03%	**0%**	**0%**
	MS	**77**	78.6	**77**	82.1	80	83	**77**	78.7	**77**	77.1	**77**	**77**	**77**	**77**	**77**	77.3
	Gap_{MS}	**0%**	2.04%	**0%**	6.21%	3.75%	7.23%	**0%**	2.16%	**0%**	0.13%	**0%**	**0%**	**0%**	**0%**	**0%**	0.39%
Ex-2	SOC	**399**	403.9	**399**	414.1	429.5	437.4	**399**	409.1	406.5	414.7	**399**	403.65	**399**	403.7	**399**	400.7
	Gap_{SOC}	**0%**	1.21%	**0%**	3.65%	7.1%	8.78%	**0%**	2.47%	1.85%	3.79%	**0%**	1.15%	**0%**	1.16%	**0%**	0.42%
	MS	**143**	150.6	**143**	152.2	160	162.3	144	146.9	**143**	144	**143**	144.6	**143**	145.09	**143**	143.2
	Gap_{MS}	**0%**	5.05%	**0%**	6.04%	10.63%	11.89%	0.69%	2.65%	**0%**	0.69%	**0%**	1.11%	**0%**	1.58%	**0%**	0.14%
Ex-3	SOC	523.25	535.12	534.6	552.99	571.6	588.21	525.5	534.92	524.6	531.98	518.1	532.15	520.6	531.93	518.1	526.97
	Gap_{SOC}	1.18%	3.37%	3.27%	6.49%	9.53%	12.09%	1.6%	3.33%	1.43%	2.8%	0.19%	2.83%	0.67%	2.79%	0.19%	1.87%
	MS	119.9	126.17	125	131.33	133.5	135.53	119.4	130.93	122	126.25	117	120.05	116	119.36	116.4	120
	Gap_{MS}	5.75%	10.44%	9.6%	13.96%	15.36%	16.62%	5.36%	13.69%	7.38%	10.5%	3.42%	5.87%	2.59%	5.33%	2.92%	5.38%
Ex-4	SOC	629	645.53	725	755.09	755.6	793.1	639.1	647.16	631.1	652.01	625.5	640.31	**622.1**	635.6	**622.1**	637.89
	Gap_{SOC}	1.1%	3.63%	14.19%	17.61%	17.67%	21.56%	2.66%	3.87%	1.43%	4.59%	0.54%	2.84%	**0%**	2.12%	**0%**	2.48%
	MS	148.9	153.5	141	157.29	159.6	165.84	**139**	149.13	138	149.8	136.4	142.63	136.4	140.73	137.6	142.44
	Gap_{MS}	6.65%	9.45%	1.42%	11.63%	12.91%	16.18%	**0%**	6.79%	−0.72%	7.21%	−1.91%	2.55%	−1.91%	1.23%	−1.02%	2.42%

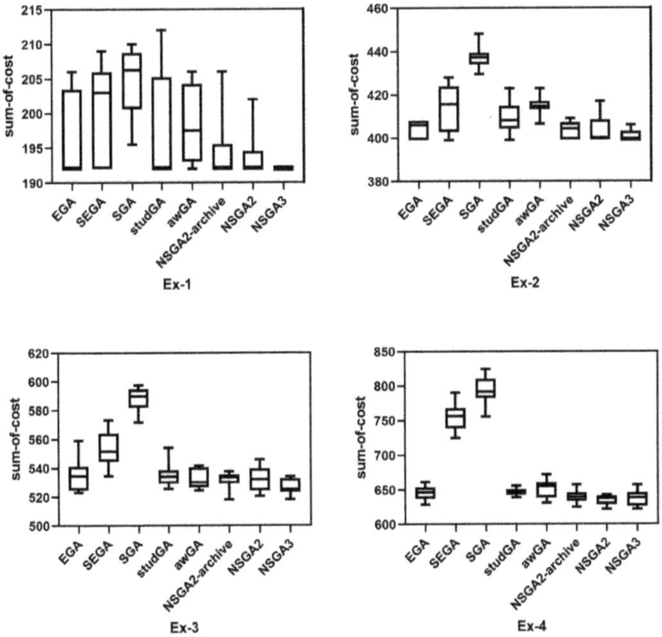

Figure 13. Box-whisker plots of sum-of-cost obtained by eight kinds of genetic algorithms on the small-scale experiments.

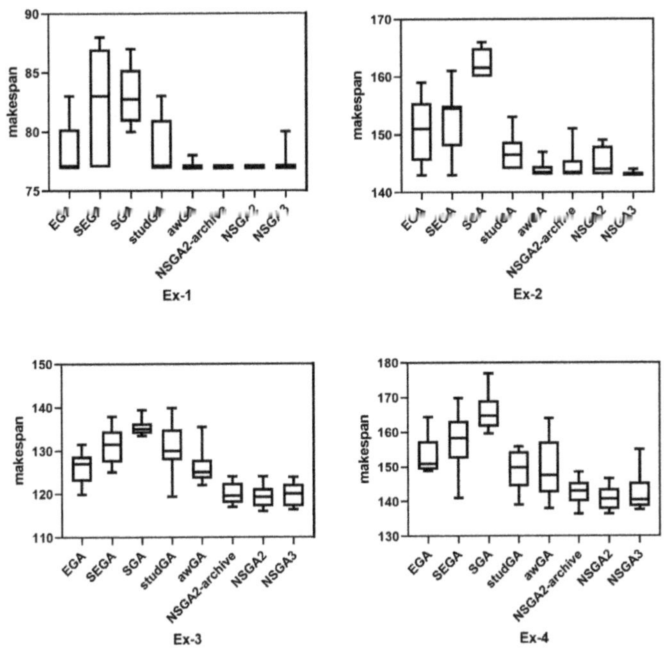

Figure 14. Box-whisker plots of makespan obtained by eight kinds of genetic algorithms on the small-scale experiments.

Table 5. Large-scale experiment results by genetic algorithms.

Index	Interation	EGA		NSGA3	
		Best	Aveg	Best	Aveg
SOC	5000	186,274	188,044	183,507	185,334
	20,000	182,712	184,076	182,074	184,494
	Gap_{SOC}	1.91%	2.11%	0.78%	0.45%
MS	5000	40,410	42,021	37,241	37,836.2
	20,000	39,552	40,455.3	37,272	37,795.7
	Gap_{MS}	2.12%	3.73%	−0.08%	0.11%

4. Conclusions

This paper proposes a two-objective integer linear programming (ILP) model of open-path multi-depot asymmetric traveling salesman problem (OP-MDATSP) for the multi-robot task assignment in an intelligent warehouse. The theoretical bound on the computational time complexity of this model is $O(n!)$. Then, we transfer the two-objective linear integer programming model to a single-objective model using the linear weight method. We solve this single-objective model on a small-scale problem using the Gurobi solver, and we compare the results with the nearest distance-based single-task assignment method. The values of the minimum sum-of-cost and makespan obtained based on the ILP model are much lower than those of the the nearest-distance-based single-task assignment (STA). After testing eight kinds of different multi-chromosome coding-based genetic algorithms in four small-scale experiments, we find that two genetic algorithms, EGA and NSGA3, perform relatively better. We use them to solve the large-scale problems. The results of the large-scale experiments show that the two genetic algorithms can solve large-scale problems well. This shows that the multi-robot task assignment method proposed in this paper has great practical significance for the intelligent warehouse to reduce its operating cost. In addition, we can adjust the weight coefficients in the objective function of this paper according to the needs of the actual scene.

For the following work, we will consider the path conflicts between robots during the task assignment of the intelligent warehouse. In addition, this paper only considers task assignment in a static state, where all task points and robot positions are determined. In the future, we will consider some unexpected scenarios, such as the dynamic change of robot state and task point state, to further verify the robustness of our method. In this paper, we assume that the robot's power is infinite and do not consider issues such as charging, which we will study in the following work.

Author Contributions: Conceptualization, formal analysis, methodology, and writing—original draft preparation by J.G.; funding acquisition, project administration, supervision, and writing—review and editing by Y.L.; data curation, investigation, resources, and software by Y.X.; validation and visualization by S.L. All authors have read and agreed to the published version of the manuscript.

Funding: Our project is supported by the Shenzhen Basic Research Program JCYJ20180507183837726 and the National Natural Science Foundation U1813206, 61977019.

Institutional Review Board Statement: Not applicable.

Informed Consent Statement: Not applicable.

Data Availability Statement: Not applicable.

Acknowledgments: We thank anyone who has provided guidance and assistance with this paper. In particular, we thank the reviewers and editors of the paper.

Conflicts of Interest: The authors declare no conflict of interest.

References

1. deKoster, R. Automated and Robotic Warehouses: Developments and Research Opportunities. *Logist. Transp.* **2018**, *38*, 33–40. [CrossRef]
2. Farinelli, A.; Boscolo, N.; Zanotto, E.; Pagello, E. Advanced approaches for multi-robot coordination in logistic scenarios. *Robot. Auton. Syst.* **2017**, *90*, 34–44. [CrossRef]
3. Wurman, R.P.; D'Andrea, R.; Mountz, M. Coordinating hundreds of cooperative, autonomous vehicles in warehouses. *AI Mag.* **2007**, *29*, 1752–1759.
4. NZanywayingoma, F.; Yang, Y. Effective task scheduling and dynamic resource optimization based on heuristic algorithms in cloud computing environment. *KSII Trans. Internet Inf. Syst.* **2017**, *11*, 5780–5802.
5. Erdmann, M.; Lozano-Perez, T. On multiple moving objects. In Proceedings of the1986 IEEE International Conference on Robotics and Automation, San Francisco, CA, USA, 7–10 April 1986; Volume 3, pp. 1419–1424. [CrossRef]
6. Wagner, G.; Choset, H.; Ayanian, N. Subdimensional Expansion and Optimal Task Reassignment. In *SOCS, Proceedings of the 5th Annual Symposium on Combinatorial, Niagara Falls, ON, Canada, 19–21 July 2012*; Association for the Advancement of Artificial Intelligence: Palo Alto, CA, USA, 2012.
7. Khamis, A.; Hussein, A.; Elmogy, A. Multi-robot task allocation: A review of the state-of-the-art. *Coop. Robot. Sens. Netw.* **2015**, *2015*, 31–51.
8. Liu, Y.; Ji, S.; Su, Z.; Guo, D. Multi-objective AGV scheduling in an automatic sorting system of an unmanned (intelligent) warehouse by using two adaptive genetic algorithms and a multi-adaptive genetic algorithm. *PLoS ONE* **2019**, *14*, e0226161. [CrossRef]
9. Pan, X.Y.; Wu, J.; Zhang, Q.W.; Lai, D.; Xie, H.L.; Zhang, C. A case study of AGV scheduling for production material handling. *Appl. Mech. Mater.* **2013**, *411*, 2351–2354. [CrossRef]
10. Zheng, Y.; Xiao, Y.; Seo, Y. A tabu search algorithm for simultaneous machine/AGV scheduling problem. *Int. J. Prod. Res.* **2014**, *52*, 5748–5763. [CrossRef]
11. Mousavi, M.; Yap, H.J.; Musa, S.N.; Tahriri, F.; Md Dawal, S.Z. Multi-objective AGV scheduling in an FMS using a hybrid of genetic algorithm and particle swarm optimization. *PLoS ONE* **2017**, *12*, e0169817. [CrossRef]
12. Su-yan, T.; Yi-fan, Z.; Li, Q.; Yong-lin, L. Survey of task allocation in multi Agent systems. *Xi Tong Gong Cheng Yu Dian Zi Ji Shu [Syst. Eng. Electron.]* **2010**, *32*, 2155–2161.
13. Axsäter, S. On the first come–first served rule in multi-echelon inventory control. *Nav. Res. Logist.* **2007**, *54*, 485–491. [CrossRef]
14. Shi, J.; Bao, Y.; Leng, F.; Yu, G. Priority-Based Balance Scheduling in Real-Time Data Warehouse. In Proceedings of the 2009 Ninth International Conference on Hybrid Intelligent Systems, Shenyang, China, 12–14 August 2009; Volume 3, pp. 301–306. [CrossRef]
15. Bolu, A.; Korçak, Ö. Adaptive task planning for multi-robot smart warehouse. *IEEE Access* **2021**, *9*, 27346–27358. [CrossRef]
16. Zhang, J.; Yang, F.; Weng, X. A building-block-based genetic algorithm for solving the robots allocation problem in a robotic mobile fulfilment system. *Math. Probl. Eng.* **2019**, *2019*, 6153848. [CrossRef]
17. Vivaldini, K.; Rocha, L.; Fróes, N.; Becker, M.; Moreira, A. Integrated tasks assignment and routing for the estimation of the optimal number of AGVS. *Int. J. Adv. Manuf. Technol.* **2015**, *82*, 719–736. [CrossRef]
18. Lenstra, J.K.; Kan, A.R. Some simple applications of the travelling salesman problem. *J. Oper. Res. Soc.* **1975**, *26*, 717–733. [CrossRef]
19. De Koster, R.; Le-Duc, T.; Roodbergen, K.J. Design and control of warehouse order picking: A literature review. *Eur. J. Oper. Res.* **2007**, *182*, 481–501. [CrossRef]
20. Theys, C.; Bräysy, O.; Dullaert, W.; Raa, B. Using a TSP heuristic for routing order pickers in warehouses. *Eur. J. Oper. Res.* **2010**, *200*, 755–763. [CrossRef]
21. Azadnia, H.A.; Taheri, S.; Ghadimi, P.; Saman, Z.M.M.; Wong, Y.K. Order batching in warehouses by minimizing total tardiness: A hybrid approach of weighted association rule mining and genetic algorithms. *Sci. World J.* **2013**, *2013*, 246578. [CrossRef]
22. Tsiropoulou, E.E.; Paruchuri, S.T.; Baras, J.S. Interest, energy and physical-aware coalition formation and resource allocation in smart IoT applications. In Proceedings of the 2017 51st Annual Conference on Information Sciences and Systems (CISS), Baltimore, MD, USA, 22–24 March 2017; pp. 1–6.
23. Benavent, E.; Martínez, A. Multi-depot multiple TSP: A polyhedral study and computational results. *Ann. Oper. Res.* **2013**, *207*, 7–25. [CrossRef]
24. Dantzig, G.; Fulkerson, R.; Johnson, S. Solution of a large-scale traveling-salesman problem. *J. Oper. Res. Soc. Am.* **1954**, *2*, 393–410. [CrossRef]
25. Odili, J.B.; Noraziah, A.; Zarina, M. A Comparative Performance Analysis of Computational Intelligence Techniques to Solve the Asymmetric Travelling Salesman Problem. *Comput. Intell. Neurosci.* **2021**, *2021*, 6625438. [CrossRef] [PubMed]
26. Bixby, B. The gurobi optimizer. *Transp. Res. Part B* **2007**, *41*, 159–178.
27. Liu, Z.; Liu, G.; Wang, H.; He, F. The linear weighting method for solving a class of non-differentiable multiobjective programming problem. In Proceedings of the 2011 International Conference on Multimedia Technology, Hangzhou, China, 26–28 July 2011; pp. 3273–3276.
28. Ronald, S.; Kirkby, S.; Eklund, P. Multi-chromosome mixed encodings for heterogeneous problems. In Proceedings of the 1997 IEEE International Conference on Evolutionary Computation (ICEC '97), Indianapolis, IN, USA, 13–16 April 1997; pp. 37–42. [CrossRef]

29. Ciesielski, V.; Scerri, P. Compound optimisation. Solving transport and routing problems with a multi-chromosome genetic algorithm. In Proceedings of the 1998 IEEE International Conference on Evolutionary Computation Proceedings. IEEE World Congress on Computational Intelligence (Cat. No. 98TH8360), Anchorage, AK, USA, 4–9 May 1998; pp. 365–370. [CrossRef]
30. Ye, D.; Liu, G.; He, B. Multi-chromosome Genetic Algorithm for Multiple Traveling Salesman Problem. *J. Syst. Simul.* **2019**, *31*, 36–42.
31. Khatib, W.; Fleming, P.J. The stud GA: A mini revolution? In Proceedings of the International Conference on Parallel Problem Solving from Nature, Amsterdam, The Netherlands, 27–30 September 1998.
32. Deb, K.; Pratap, A.; Agarwal, S.; Meyarivan, T. A fast and elitist multiobjective genetic algorithm: NSGA-II. *IEEE Trans. Evol. Comput.* **2002**, *6*, 182–197. [CrossRef]
33. Deb, K.; Jain, H. An Evolutionary Many-Objective Optimization Algorithm Using Reference-Point-Based Nondominated Sorting Approach, Part I: Solving Problems with Box Constraints. *IEEE Trans. Evol. Comput.* **2014**, *18*, 577–601. [CrossRef]

MDPI AG
Grosspeteranlage 5
4052 Basel
Switzerland
Tel.: +41 61 683 77 34

Applied Sciences Editorial Office
E-mail: applsci@mdpi.com
www.mdpi.com/journal/applsci